More praise for

THE FEW AND THE PROUD

"Marine D.I.s have an awesome responsibility: that of transforming an unorganized, unmotivated, sloppy, unfit civilian into a hard charg'n, low crawl'n, highly disciplined, hard as nails United States Marine. Larry Smith's interviews reveal how this is accomplished. The drive, determination, devotion to duty and professionalism of D.I.s ensure our Corps of Marines remains the best of the best, a true Band of Brothers." —H. C. "Barney" Barnum Jr., Colonel of Marines (Ret.), Medal of Honor Recipient

"Distinguished author Larry Smith turns his reporter's eye to the world's most renowned training facility, the place where they make Marines, to write one of the most engaging nonfiction books you'll ever read. Rarely will a reader find truth, detail, and action woven together so skillfully, so compellingly. Smith is a true storyteller. Parris Island comes alive in the energizing words of those who've lived it: you smell the air, taste the salt, feel the sand." —Walter Anderson

"Smith captures the men and women who are the essence of the Marine Corps." —*The Pilot*, North Carolina

"A multifaceted, personal history of the Marine Corps." —*Publishers Weekly*

"Smith's collection of interviews illuminates one of the U.S. Marine Corps' institutional legends, the DI." —*Booklist*

"A superb job of describing how the Corps creates a brand of warrior whose very mention puts the fear of God into their enemies . . . first-hand accounts from Marine drill sergeants, whose tales include everything from training recruits to the hell of combat." —Military Book Club, as quoted in *San Diego Reader*

"Culminating in the experiences of 28 Marines from Parris Island to MCRCD San Diego, *The Few and the Proud* brings to life the motivations drill instructors use to push themselves, and their recruits, to the limits, giving them some of the toughest training and richest traditions in the military." —Corporal Brian Kester, *The Boot*

"Together, their accounts contribute to a better understanding of the DI experience, ethos and methods."

—Thomas R. Kailbourn, *Military Trader*

From the sands of Iwo Jima to the deserts of Iraq

THE FEW
AND THE
PROUD

*Marine Corps
Drill Instructors
in Their Own
Words*

LARRY SMITH

W. W. NORTON & COMPANY

New York · London

For information about permission to reproduce selections from this book, write to Permissions,
W. W. Norton & Company, Inc., 500 Fifth Avenue, New York, NY 10110

Manufacturing by Courier Westford
Book design by Charlotte Staub
Production manager: Anna Oler

Library of Congress Cataloging-in-Publication Data

Smith, Larry, 1940–
The few and the proud : Marine Corps drill instructors in their own
words / Larry Smith.
p. cm.
ISBN-13: 978-0-393-06044-7 (hardcover)
ISBN-10: 0-393-06044-6 (hardcover)
1. United States. Marine Corps—Non-commissioned officers—Interviews. 2. Basic training
(Military education)—United States. 3. United States. Marine Corps—Military life. I. Title.
VE24.S58 2006
359.9'650973—dc22   2006001108

ISBN 978-0-393-32992-6 pbk.

W. W. Norton & Company, Inc.
500 Fifth Avenue, New York, N.Y. 10110
www.wwnorton.com

W. W. Norton & Company Ltd.
Castle House, 75/76 Wells Street, London W1T 3QT

1 2 3 4 5 6 7 8 9 0

**FOR S/SGT.
EDDIE ADAMS**

*1933–2004
Semper Fi, Mac*

CONTENTS

Come on, you sons of bitches—
do you want to live forever?

—attributed to Gunnery
Sergeant Daniel Daly, USMC,
Belleau Wood, June 1918.
Two-time recipient,
Medal of Honor

ACKNOWLEDGMENTS

Any flaws found in this book are, as Mark Twain might have said, the exclusive property of the author. All the good parts were built on strong contributions of several individuals, notably Major Kenneth White (USMC Ret.), who was in charge of public affairs at the Marine Corps Recruit Depot at Parris Island from August 2002 until he retired in early 2006. He was endlessly patient with an ignorant civilian, made splendid suggestions, and dug up phone numbers and facts that greatly enhanced the enterprise. The book itself grew out of a suggestion by Walter Anderson, the chairman and CEO of *Parade* magazine, who was designated in November of 2005 as a Marine for Life. I cannot thank him sufficiently for all he has done. I also want to thank Lee Kravitz for expert guidance. Col. Mike Malachowsky, chief of staff at Parris Island prior to joining the United States Special Operations Command in Tampa in September of 2004, was kind enough to read parts of the manuscript and offer perceptive suggestions. He also led me to Robert Mastrion, a distinguished former Marine who suggested some excellent interview subjects. Others I wish to thank are Vic Ditchkoff, president of the Drill Instructors Association, Dr. Stephen Wise, director of the museum at Parris Island, and Lt. Anthony Delsignore, deputy director of public affairs at the Marine Corps Recruit Depot in San Diego. Book titles may seem obvious once they are in place but this is not always so. That was the case with this book, and its fine title ultimately came from my wife, Dorothea. I also thank Carole Smith

Strasser in El Cajon, California, Jean Fujisaki and Robert Nelson in La Jolla, Cpl. Brian Kester of MCRD Parris Island, the incomparable researcher Lou Leventhal, the technical wizard Jonathan Au, Miriam Lorentzen, and Ponchitta Pierce. Finally, Tom Mayer of Norton was endlessly courteous, patient, and, best, extremely competent. Thank you all.

INTRODUCTION

PARRIS ISLAND AND SAN DIEGO

Where the Marine Corps Begins

The Continental ship *Providence*, now lying at Boston, is bound on a short cruise, immediately; a few good men are wanted to make up her complement.
—Marine Captain William Jones,
Providence Gazette, March 20, 1779

Certain place names in the United States, such as Gettysburg, West Point, and the Little Bighorn, evoke immediate recognition. Parris Island, the Marine Corps boot camp on the South Carolina coast, is one of these. It is four miles long, three miles wide, and known as a hard place, made special by sand fleas, stultifying heat, and six hundred fearsome, sometimes terrifying, drill instructors. Nearly everyone who joins the Marine Corps from east of the Mississippi undergoes basic training at Parris Island. Those who join from west of the Mississippi are sent to the Marine Corps Recruit Depot (MCRD) in San Diego. No one ever forgets basic training in the Marine Corps. More than two million young men and women have survived training at the two bases, and few, if any, were unaffected by the experience. Marine Corps recruiters maintain, "The change is forever." They rarely get an argument.

This book is about the experience of training recruits in the Marine Corps, as seen for the most part by its drill instructors, both active and retired. Men who underwent training as Marines have fought in every American conflict since the Revolution. The D.I.s interviewed for this book helped to train the toughest element of one of the world's dominant armed forces. As the following chapters will show, any history of Parris Island, San Diego, and their drill instructors is also a history of the Marine Corps itself and, to some extent, of the United States.

The French were the first Europeans to arrive at Parris Island when they landed in 1562. They built an outpost called Charlesfort, but abandoned it only a year later. The place was taken over by the Spanish in 1566. Later the English occupied the island as they expanded their influence in America. The nearby city of Beaufort was founded in 1711, four years before the island was purchased by a Colonel Alexander Parris and given its current name. From the time of the Revolution until after the Civil War, Parris Island was mainly plantation country. In the 1890s, a huge drydock was built and a naval station was established there and the place became known as Port Royal. The drydock was moved to Charleston in 1909 and Parris Island became a school for Marine Corps officers. Two years later, in 1911, the Navy built disciplinary barracks for over three hundred prisoners. The island was officially designated a Marine Corps Recruit Depot on November 1, 1915, though everyone arrived by boat until a causeway and bridge to the mainland were completed in 1929. It was on Parris Island in 1917 and 1918 that most of the recruits who were to fight in World War I completed their training, and it has been one of the most fertile training grounds for the Marine Corps ever since.

"The American flag flew on the West Coast for the first time in 1843, at the end of the Mexican American war, when a detachment of Marines came ashore, raised the flag in San Diego's Old Town, and immediately sailed back out," says Parker Jackson, an unofficial historian of the Recruit Depot in San Diego. "They did not return until 1913, under the command of Col. Joseph H. Pendleton. He was on his way to Central America to deal with potential political unrest, which did not materialize. They camped at North Island in San Diego harbor before they went south.

"The next year, the Marines were invited to occupy an unused space that was part of an international exposition in Balboa Park in San Diego. They served as an honorary military presence for the next two years, then leased some vacant buildings. Some years earlier Congress had purchased 250 acres in a low-lying tidal marsh, near what is today the San Diego airport, for $250,000. The City of San Diego donated an additional five hundred acres. The Navy even contracted with a well-known architect, Bertram Goodhue, to create a site plan and design some buildings for a base. The only problem was that the area, called Dutch Flats, was uninhabitable." Eventually, says Jackson, the flats were filled in with dredging from San Diego Harbor. "The Depot today is built on twenty to thirty-five feet of landfill and goes all the way to the harbor. In the years during and after World War I, the airport was an insignificant little spot on the Pacific Highway, which back then was called Atlantic Avenue. Building 1, the Receiving Barracks, and Barracks 2, 3, 4, and 5 were dedicated in 1921."

Training at the Marine Corps Recruit Depot in San Diego began in 1923 after Colonel Pendleton and a congressman, Rep. William Kettner, selected the area for a Marine base to prepare recruits for expeditionary duty. Previously, Marines were trained at Mare Island Navy Yard in Vallejo, north of San Francisco. Prior to World War II, about 270 recruits per month were trained by 33 drill instructors. Today more than half of all men, about 20,000 annually, are trained at San Diego. All women recruits are sent to Parris Island. The Marine Corps is consolidated into seventeen bases around the world, with five major West Coast installations in southern California and Arizona.

The function of the Marine Corps is officially described as organizing, training, and equipping the Fleet Marine Force (FMF) for seizing or defending advance naval bases, providing security for naval vessels, stations, and bases, coordinating with the Army and Air Force amphibious operations relating to landing forces, and performing other duties at the direction of the President. But Col. Mike Malachowsky, chief of staff at the Recruit Depot at Parris Island in 2004, declared: "The Marine Corps wins battles; the Army wins wars. The function of the Marine Corps is to locate, close with, and kill the enemy." With that goal in mind, the Corps has long dictated that

"Every Marine is a rifleman," and every Marine, whether he or she becomes a cook, a clerk, a helicopter pilot, or an embassy guard, practices accordingly.

From the time of its creation, the Corps has distinguished itself in combat. The Civil War saw 148 men killed in action on both sides as half of its officers and two thirds of its lieutenants resigned to join the Confederacy. During that war, the Corps tripled in size to a modest 4,167 men. Throughout the nineteenth century, Marines fought all over the world, from Mexico to the Caribbean to the Philippines to China, though they did not become a substantial force until World War I, when the number of enlisted men and officers grew from 15,000 to 76,000, half of whom went to France. These Marines fought famously against the Germans at Belleau Wood on June 6, 1918; 1,000 were killed or wounded the first day of the battle. It was here, it is said, that the enemy first called them *Teufelhunden*, or Devil Dogs. In August of 1918, the first women joined the Corps; three hundred signed up for clerical duty in Washington, D.C., "freeing men to fight." They were called Marinettes.

By 1940, with World War II under way in Europe, the Corps had grown to 30,000; by war's end the total number of Marines would peak at 485,000, including 20,000 blacks and 20,000 women. More than 19,000 would be killed and 67,000 were to be wounded in three and a half years of fighting, mostly against the Japanese in the Pacific. Here the Marines distinguished themselves again, in places with names that are synonymous with valor and American military might, such as Guadalcanal, New Guinea, Tarawa, Saipan, and Okinawa. Perhaps the signature battle of the Marine Corps to this day remains the invasion of Iwo Jima, which began in the early morning of February 19, 1945. A total of 71,000 men from three Marine divisions went ashore during six nightmarish weeks. More than 6,300 Marines were killed and 19,000 were wounded, while 23,000 Japanese soldiers died. Twenty-six Medals of Honor were awarded, according to the Congressional Medal of Honor Society. On the fourth day of the battle, five Marines and a Navy corpsman raised the flag on the summit of Iwo's Mount Suribachi, replacing a smaller one that had flown there first. Joseph Rosenthal of the Associated Press was there to immortalize the moment, and his award-winning photograph became the enduring image of the war and the model for the Marine Corps Memorial in Washington, D.C.

After the war, the Marines were the most famous and most well-respected division of the military. On the orders of President Harry Truman, both blacks and women began training at Parris Island in 1948, but this did not stop the Corps from shrinking as Marines received their discharge papers and returned home after the war. By the summer of 1950, when the Korean War began, the number of active Marines was down to 75,000, though 100,000 new recruits came on board during the conflict, raising the total to 175,000.

Then came April 8, 1956, when a Parris Island drill instructor named Staff Sgt. Matthew C. McKeon led Platoon 71 on a night march into a stream called Ribbon Creek. The platoon panicked when it got caught in an unusually strong tidal current and six men drowned. The incident, and the publicity from McKeon's subsequent court-martial, almost wrecked the Marine Corps. The country was outraged at what it viewed as vicious training practices and negligence on the part of the drill instructors. In the eyes of many, the Corps had become a brutal place where inhumane drill instructors battered the nation's young men.

Sgt. McKeon went before a military court facing charges tantamount to homicide and, brilliantly defended by the New York attorney Emile Zola Berman, was found guilty on August 3, 1956, of simple negligence and drinking on duty. Instead of possible jail time and a dishonorable discharge, McKeon received a bad conduct discharge, forfeited $30 a month for nine months while confined to hard labor, and was reduced to the rank of private. The following October, the Secretary of the Navy voided the bad conduct discharge and the fine, cut the period of confinement to three months, and let the reduction to private stand. McKeon was finally freed on October 19 and returned to duty. Through Zola's management, the trial focused on the propriety of the rationale and practices of Marine Corps training as much as it did on McKeon's responsibility for the recruits' death.

Significant changes in recruit training followed, but it took awhile and, at the same time, lapses were inevitable. Recruits die from time to time through natural or accidental causes, such as heart or respiratory failures. A second training failure, known as the McClure incident, occurred in 1976 at the Marine Corps Recruit Depot in San Diego when a twenty-year-old trainee, Private Lynn E. McClure of Lufkin, Texas, died from excessive beating with a pugil stick. (Pugil sticks have boxing-type gloves on both ends and are used to teach assault and defense simulating close combat with rifle and bayonet.) Two staff sergeants were court-martialed. One was acquitted; the other was convicted of dereliction and reprimanded. There was also an incident at Parris Island where a drill instructor accidentally shot a recruit in the hand with an M-16. This understandably led to even greater oversight in recruit training. The Standard Operating Procedure (SOP) outlining the care and treatment of recruits grew from a few pages to a book-length guide as it spelled out every aspect of the relationship between the drill instructor and the recruit. In today's Marine Corps, no physical contact by the D.I. is permitted, and a D.I. may be relieved for cause, fined, and reduced in rank with his career effectively stalled if he or she is found to have violated the SOP.

The training course for drill instructors lasts twelve weeks. They serve three years and get special duty pay and a ribbon. They also may subsequently perform security guard duty in embassies and at other sites. On the drill field, they not infrequently work up to 120 hours a week. The drill instructor is the only person in the Corps who wears the famous flat-brimmed Smokey, or the preferred "campaign hat." It is the only military hat or "cover" that can be worn indoors.

———

What makes Parris Island, San Diego, and the Marine Corps special? It has been said that America doesn't *need* the Marine Corps; America *wants* the Marine Corps. Those who join the Corps undergo a transformation in which they relinquish individuality for the sake of the team, the squad, and the platoon. They acquire discipline and the skills to respond instantly to orders while maintaining the ability to think and lead. Sgt. Major Mike Mervosh, whose story is told in Chapter 2, went from a sergeant commanding a machinegun squad on Iwo Jima to company commander because all of his superior officers were killed or wounded. The Corps emphasizes to recruits what it calls its "core values"—honor, courage, and commitment. Great stress is placed on individual integrity, presented as keeping one's word. The traditions of the Corps, as well as the name of the Commandant, are drilled into the heads of recruits. Recruits also learn about Marine Medal of Honor recipients, whom the Corps regularly venerates, and the outstanding performance by the Marines in the Korea and Vietnam conflicts. Indeed, stations at the Crucible, which is the ultimate test undergone by all recruits near the end of their twelve weeks of training, are named for Medal of Honor recipients. Drill instructors make it a point to tell the story of each man to the recruits as they pass through.

Whereas it was not unusual at mid-century for young men, many of them dropouts, to go straight from jail into the Marine Corps, today's Marines must have clean records and they must be high school graduates. Approximately 36,000 young men are recruited annually for a four-year enlistment period. Of that number only 4,000 are kept in the Corps after their enlistment period is up. "You have to compete for boat spaces," says Major Kenneth White, the public affairs officer at Parris Island, who was to retire in early 2006. "The Marines are constantly sending back into society responsible, upstanding patriots who will serve their communities and continue to carry the torch for the Corps."

In addition, up to 2,200 women are trained annually in an all-female Marine battalion at Parris Island. Why do they join? Mary Sue League, who retired a lieutenant colonel in 1985, declares: "I think there's as many reasons for women to join as there are for men. They don't have money to go to college, they may not be able to find a job locally, they want to further their education, and they want to serve their country. That's paramount. I had six brothers in World War II. My whole family was very patriotic. And, yes, I saw *Sands of Iwo Jima*, too."

Women have not always been fully accepted members of the Marine Corps. Whereas they once trained as clerical workers and hostesses, female Marines at Parris Island began shooting for score at the rifle range in 1985. Training for men and women became uniform in 1988. Women today fly all Marine Corps aircraft—except for attack Cobras, because females are excluded from combat in the Marines, as they are in all the military services. However, in a battle setting such as Iraq, where there are no front lines, women may be expected to confront live fire. Everyone is conscious of this and all Marines, whether male or female, prepare as if they were going to war.

Like women, African Americans had to overcome prejudice in order to win equal treatment. The first black recruits joined the Corps in 1942, but they were trained at segregated Montford Point, part of Camp Lejeune in North Carolina. Originally, the Marines used black recruits as service troops. They acted as stewards and served in ammo depot companies for the duration of the war. Twenty-three thousand trained at Montford Point. In 1949, when Harry Truman integrated the military, black and women Marines began training at Parris Island. By 1998, black Marines made up 7 percent of the Corps's officer complement, and 17 percent of the enlisted personnel.

While training components are fundamentally unchanged to this day, training itself is much more flexible. Recruits arrive by bus, whether at Parris Island or the San Diego depot, usually after dark, when they are more likely to be disoriented. Their thirteen weeks (one week for Forming) of training begins when a drill instructor wearing a hard-brimmed campaign hat comes in yelling that he will give them only seconds to get off "his" bus and line up on the Yellow Footprints painted on the street outside; otherwise, they face unmentionable peril.

From their first moments at Parris Island or the MCRD in San Diego, the recruits will learn discipline and immediate obedience to orders. They also will learn to march, to move through water with packs on, to rappel from sixty-foot towers; they will practice hand-to-hand combat, and fight with the bayonet—simulated by use of the pugil stick. They must shoot and qualify with the M-16 rifle, handle gas masks, and solve both the Confidence and the Obstacle Courses. The twelve-week training cycle culminates in the Crucible, a fifty-four-hour challenge that happens in the eleventh week of training and is one of the final tasks a recruit must undergo before he or she can be called a Marine. The platoon is awakened at 1:30 A.M., marched six miles, and broken up into squads that must face a variety of tasks, called events, which can only be accomplished through teamwork. Each squad must designate a leader and then solve the problems inherent in each event. They proceed through a number of events simulating combat situations, such as tactical assault, moving one, then two, wounded Marines over a mile, transporting an awkward, heavy dummy over a series of obstacles, combat assault resupply, combat endurance, and night infiltration with a simulated live-fire exercise. They will get only four hours of sleep and limited rations. At the end, the platoon gets up at 3:00 A.M. and hikes nine miles back to barracks (throughout the Crucible, the recruits will have marched over forty-eight miles). If successful, the recruits receive the Marine Corps emblem, the Eagle, Globe and Anchor, which dates from 1868, on Thursday at 1:00 P.M. at the Parade Deck. This is followed by graduation at 9:00 A.M. Friday. Here the troops are arrayed in formation, and parents and loved ones sit in the bleachers and watch as they are formally invested as United States Marines. Recruits first encountered the Crucible on December 12–14, 1996.

"When the senior D.I. shook my hand," a young recruit declares in the documentary *Marine: Earning the Title*, "it was the first time he called me Marine. Receiving the Eagle, Globe and Anchor was the best feeling I ever felt in my life. I could feel the Marine Corps being tattooed on my soul forever. It's worth every inch of blood, sweat, and tears you put into it. It's a pride you can't buy. You have to earn it."

A great many civilians, including this writer, also saw *Sands of Iwo Jima* and heard about how tough were those China, Nicaragua, World War II, and Chosin Reservoir Marines, many of them high school dropouts, many a half-step from jail on going in. They call them the Old Breed to this day. They were rough and tough and ready. Their record speaks for itself.

Seventy thousand Marines are stationed in Iraq as these words are being written. The question in my mind, in setting out to explore recruit training, was this: How do today's Marines stack up? As a recruit today, you can't have been in jail, you have to be a high school graduate, you get to use Skin So Soft to ward off the sand fleas, and you run, for the most part, in sneakers. Military vans containing corpsmen and ice water follow close on the heels of platoon runs. Despite all that, the Marines remain the proudest, toughest, most disciplined soldiers in the United States military. The following pages tell what I learned from the drill instructors who train them.

—Larry Smith
January 2006
Norwalk, Connecticut

THE FEW
AND THE
PROUD

PART ONE

THE OLD BREED

World War II, Korea,
and Vietnam

WHILE THE MARINES distinguished themselves over the decades fighting the Barbary pirates on the coast of Africa, confronting the Boxer Rebellion in China around the turn of the century, and, later, fighting in the Philippines and Central America, they really made a name for themselves fighting in World War II, Korea, and Vietnam. They fought nearly all their battles in World War II in the Pacific, invading, seizing, and then holding on for dear life to islands such as Guadalcanal, Tarawa, Roi-Namur, Tinian, Saipan, Iwo Jima, and, finally, Okinawa. Their deeds and sacrifices in these settings are legendary.

Dedication, discipline, and esprit de corps, as well as the relatively small size of the force, set the Marines apart from the Army and the Navy. The basic training experience, as these pages will show, is central to the making of a Marine, and it is something they never forget.

The Marine Corps has always been extremely savvy when it comes to public relations, and has always found ways to distinguish itself as the most appealing of all branches of the military. It has found glamour in toughness, and fashioned an appeal accordingly, as the story of Chuck Taliano indicates.

While the "Old Breed" more properly refers to the Marines who fought in World War I and in the Caribbean during the 1930s, there's always an Old Breed, and it's likely to be the generation just before yours. Taliano, star of a recruiting poster, and Mike Mervosh, an Iwo Jima veteran, Bill Paxton, Ed Walls, and Dave Robles, old-school drill instructors, and Robert Mastrion, who was court-martialed twice for fighting and yet went on to earn the Silver Star for service in Vietnam, represent today's version of the Old Breed, stand-up guys who went in willing to perform what was asked of them, men who see changes in today's Marines that may make them uncomfortable but do not diminish their faith in the Corps.

CHAPTER ONE

CHUCK TALIANO

Poster Marine
1964–1968

Let no man's ghost say,
"If you'd only done your job."

A chance photo of stern, unsmiling Chuck Taliano, taken in April of 1968 when he was a drill instructor, was widely circulated in a major Marine Corps recruiting campaign that began in the fall of 1971. The recruit has never been identified. This photo of Taliano, holding the poster, was taken in 2004, just before he turned fifty-nine. (*Photograph by Larry Smith*)

Because of a recruiting poster first published in 1971, the face of Chuck Taliano, who operates the museum store on Parris Island, is known to almost everyone with a nodding acquaintance of the Marine Corps. Taliano was having a drink with another former drill instructor, Lou Wasirick, and a couple of others during a D.I. reunion on Parris Island in the spring of 2004 when they bumped into a guy named Mervosh, known as Iron Mike. He acquired the nickname over thirty-five years in the Marine Corps, rising from corporal to company commander of his unit on Iwo Jima after all his superior officers had been killed or wounded. Mervosh served nineteen years as a sergeant major, fighting in Korea and Vietnam as well. Mervosh turned to introduce Taliano to a stranger. "Look at this little son of a seabiscuit," he said. "I've got more time in the crapper as a corporal than he does for one hitch, and he's got just about the most recognizable face in the whole Marine Corps."

"I was born in Cleveland in May of 1945," Chuck Taliano said, "and went into the Marine Corps in February of 1964. Growing up, I always felt strongly about the military, even though I came from a family that kind of demanded that you have a college education. I enrolled in community college in Cleveland but from the time I was a junior in high school I had already visited most of the military recruiters. I have to say I was most impressed with the Marine recruiter.

"The day the president, John F. Kennedy, was assassinated, November 22, 1963, I went from my junior college class to see the Marine recruiter. I had no thought of enlisting at that point: He was just like a Rock of Gibraltar in my mind, and I needed to talk to someone.

"He was a good salesman. He said: 'The president has been assassinated. The country is in peril. Without your enlistment, the country may fall to the communists.'

"There were seven of us who said, 'We're going to go.' Now this was interesting because, if the country was imminently going to fall to the communists, why were we not going to leave for boot camp until the following February? Simple: He had a quota to fill.

"The draft was in effect then and I thought if I was going to serve my country I'd better serve it in the Marines because of their reputation. I had relatives who had served in the Navy, the Coast Guard, and the Army. Most of them served in World War II. I had an uncle go through five major campaigns in the Pacific. He would not talk about it.

"I arrived at Parris Island in the dark on Valentine's Day 1964. I said, What the hell am I doing here? Why did I do this? The recruiter didn't lie. He said it was going to be challenging, demanding. He said it would be the toughest accomplishment we'd have to go through in our lifetime but it would be worth it. He was right: It was a real challenge, every day. I kept waiting for the left shoe to drop, for things to lighten up, for the drill instructor to say, 'Okay, We're all buddies now, let's train.' But instead it just got tighter every day.

"We had thirteen weeks of training and then in April of 1964 I went to Camp Lejeune for infantry training, reporting after a leave to the Second Marine Division. I didn't serve in Vietnam. We deployed to the Dominican Republic in May of 1965, returning sometime in June. It was a police action, much like the recent situation in Haiti.

"We came back to Jacksonville, North Carolina, where Camp Lejeune is situated, but this wasn't my kind of town. A lot of the younger guys liked to go in and get drunk and raise hell, which did not seem purposeful, in my mind. Before we left for Santo Domingo, I was volunteering for everything, to go anyplace, just to get out of there. In fact, I had been accepted to embassy school, but my orders were canceled when we deployed, which was pretty typical back then. And when we arrived back aboard ship, after the Dominican Republic, the Gulf of Tonkin incident occurred in August of 1964. We had a morning muster, information would be passed around and they would have requests for people to go to the Western Pacific, Westpac. I don't think they even used the acronym then. But I used to joke with the guys at lunchtime. I'd say, 'Do you know what you're volunteering for? Western Pacific is not a railroad. There's bombs going off.'

"Even so, I was one of the first to volunteer. It was never my MOS [Military Occupational Specialty]. My job was 34-1, which was in supply. I was in artillery. I went to see the battery first sergeant and asked to change my MOS. But he wouldn't do it. Then I read in the base newspaper that the D.I. screening board from Parris Island was going to be aboard the base and here were the qualifications. If you were interested, see your company first sergeant. So I did, and he said, 'I'm going to get the old man to recommend that you go because I'm just tired of you in my office every day volunteeering for stuff.'

"So I went before the screening board. Then at the morning muster next day there was a quota for two with my MOS for Vietnam. Westpac. So a buddy and I went to the first sergeant's office. I got in there and the first sergeant looked at me and said, 'I'm sick and tired of this. You're not going anywhere because you're on administrative hold.' Usually when you hear these kinds of words, it's like office hours: You're in trouble.

"I was a young corporal. I said, 'First Sergeant, I don't understand.' He said, 'Where were you yesterday?'

" 'I was over in supply.'

" 'Where did you go meanwhile?'

" 'I went before the D.I. screening board.'

"Then he said, 'I don't know why, but they recommended you to drill instructor school, and until such time as they make a determination to give the orders or not, you can't go any place. You are on administrative hold.' Consequently, I received orders for drill instructor school.

"I went before the board in November of 1965, the orders came through, I went on leave in December and reported for D.I. school in February of '66, graduating in April. I served two years as a drill instructor, retiring in May of 1968.

"Being a D.I., other than being a father, was the most rewarding thing in my life," says Taliano. Most D.I.s didn't realize how great their impact had been. "I heard from some of these former recruits who got in touch, and they'd say, 'I don't know what I would have done. My life could have been a shambles. I had no purpose, I would never have survived Vietnam if it hadn't been for you and for my training.' That's awe-inspiring. So being a D.I. is the toughest job that you love to hate, or you hate to love because the hours were difficult, the pressure was constant, and we worked a hundred hours a week. There was a period during the Tet Offensive in 1968 when there were not enough drill instructors to go around and we were working two-man teams—we never had a day off.

"This was ten years after Ribbon Creek and we never went down there to the water again. Training was very controlled but you didn't feel like someone was looking over your shoulder. We were stern with the recruits. The pressure after Vietnam started was intense, much of it self-inflicted. We knew guys were going to fight and get killed. There used to be a saying: 'Let no man's ghost say, "If you'd only done your job." ' You'd look at a recruit and you just had a feeling that he was not going to be around in another year and what you didn't want to do was to feel that you were responsible because you didn't train him to think fast enough or move fast enough.

"The real emphasis here at Parris Island was discipline. That was the most important thing. Hygiene, how to wear the uniform were part of discipline as well. Every Marine is a rifleman first, enlisted or officer, and every Marine goes through combat training. Those who don't have the infantry MOS won't go through the advanced course but, no matter what your rank is, you're always ready to step in and take the next rank.

"Drill instructor duty along with recruiting duty are the two most demanding jobs. Recruiting duty can be considered among the most demanding," says Taliano, because you work long hours and the job is thankless. Nevertheless, he felt being a D.I. carried more responsibility. "As a D.I., you want them to be perfect, you want them to maintain the tradition." It is a stressful job for someone barely of college age. Says Taliano, "You had to be a young man to be a drill instructor. I turned twenty-one years old three days before I graduated from D.I. school; that was one of the requirements. You had to be twenty-one. In fact, I had recruits who were older than I was.

"The training went from twelve weeks to ten then to eight. They even tried a couple six-week cycles but it was a disaster. The extra time allowed you to get them in better physical shape and in a better state of discipline. You had to give them an instant willingness to obey orders. In war, you haven't got time to debate the issue, you've just got to do it. We lost 14,000 Marines in two years in Vietnam. From 1967, '68, right around Tet, we couldn't produce enough Marines. We were running 113 men per platoon.

"As time went on we'd get letters from Marines we'd trained, saying, 'How's the new herd, the new hogs?' I don't know if I should use that language but that's what we used to refer to them as. 'Are they snapping to? Are they as good as we turned out to be? We get to eat pogey bait [candy bars] here [in Vietnam]. They're not as demanding as you were.' "

Taliano says it was good to hear from Marines he had trained in boot camp. Many wrote regularly, but the news was not always good. "You'd get a letter when they were on their way home for leave, and then you'd get another letter from California where they're in training for Vietnam. Then about a month later you'd get some news that would just chew you up: Private Number So and So, we couldn't find enough to send home yesterday, or Private So and So lost both of his legs. I'd read these in the D.I. house, then I'd go out and see some kid in a minor infraction and I'd be all over him." Taliano knew that chewing the kid out would not save his life, but he also knew discipline was the only way that kid would have a chance. Says Taliano, "I finally got to the point where I wouldn't open those letters, certainly not when I was on duty.

"I still remember a lot of those recruits whose names are on that wall in Washington and I've been in touch with several others, met them personally, where they've phoned me or e-mailed me. Lou and I have had a lot of letters saying, 'Thank you for being as tough as you were. It saved my life.' That makes it all worthwhile.

"The recruiting poster picture was taken at the end of April 1968. What happened was a Marine reservist was writing a book about boot camp, later published as *The Marine Machine: The Making of a U.S. Marine,* by William Mares, published by Doubleday. The author was an accomplished photographer, among other things, and he was taking photos at random for the book. As I understand it, the terms of his arrangement with the Marine Corps were that it got to screen all the

We don't promise you a rose garden

THE MARINES ARE LOOKING FOR A FEW GOOD MEN.

(Courtesy of the United States Marine Corps)

photos and approve them for use in the book. The Corps would also have the freedom to use his photos any way it wanted. And so it happened to be that my photo, among others, was selected for a recruiting campaign. And I had been out of the Marine Corps since May of '68, but that photo and that campaign did not break until the fall of 1971.

"That recruit campaign was called 'The Marines Are Looking for a Few Good Men.' This goes back to the very origin of the Marine Corps with that captain who put out that he was looking for "a few good men" to serve aboard a ship to go take on the Barbary pirates.

" 'We Don't Promise You a Rose Garden' was one poster, one theme in that overall campaign. I had two drill instructor classmates who also appeared on posters in that series. Some of the stills, which I still have, were used in a TV commercial. Back in those days there were public service announcements of four or five minutes for the various branches of military service. They must have cost millions. They aired forever and ever, usually at o Dark Hundred [meaning midnight or later], when they would appeal to young men and women who were up at that hour.

"Lynn Anderson's song 'I Never Promised You a Rose Garden' was sort of the theme for that one particular recruiting cycle and my photo was part of that. I have a copy of the commercial."

Aside from the recognition, did Taliano profit from the poster, which has sold in the millions? "That image could have been any one of us. On any given day Lou was doing it, I was doing it, just being in the right place at the right time. It all belongs to the Marines. When I got out in '68, I went back home to Cleveland and got into retail for a couple of years and then went into book publishing and stayed there thirty years, mostly in New York. I got transferred back to Ohio and I worked in Columbus and then I moved back east, lived in New Jersey. I got my start with Random House, worked with Simon & Schuster, Putnam, Doubleday. I was in sales and marketing, then got into management. I retired two years ago from publishing, in December 2002.

"The museum shop already existed and I'd served on the board of directors. I don't own it. I serve at the pleasure of the board of directors of the museum and the historical society. The shop helps support the museum. It's a government museum.

"The poster when it came out had an immediate impact. Remember, this was 1971, and I had got out in 1968. The Vietnam war was grinding down. I was at Random House and my wife sent a copy of it to my boss, the sales manager. He had a slide made of it and showed it to everybody at a sales meeting. From that day forward it just kind of spread around. I'm told by the Marine Corps Recruiting Command and I have heard from the J. Walter Thompson Agency that this was the most successful advertising campaign the Marine Corps had to that date. It lasted to 1984. I've also been told the Thompson agency used it to recruit corporate clients. I'm hoping to meet with some Thompson people to get more history on this. J. Walter Thompson has been the Marine Corps's only ad agency from Day One, dating back probably to right after World War II.

"The first time I came back from Parris Island, I signed 500 posters in three and a half hours, and they put a picture of the poster at the back of the program at the Marine Corps birthday ball in Philadelphia, for our 225th anniversary. Then they put a spotlight on me and said: He's right there. Well, there was a mad rush. I had been divorced, and I had a date, but we never had a chance to dance because I signed posters from nine til one o'clock in the morning."

Taliano's friend Lou Wasirick, of Metuchen, New Jersey, who saw extensive combat during a thirteen-month tour in Vietnam, had been a fellow drill instructor at Parris Island. "'I was aide de camp to Chuck when we went to Philly for the Marine Corps birthday party," he recalled. "We hardly got there when they saw who he was and they all started to line up. They wanted to see the blood vessel in his neck pulsing out. Then, around 10:00 P.M., the majors and the captains started coming in. Then about midnight, the generals came in, wanting signatures. It was amazing. Another time we were at Parris Island and two Harrier jets came flying in. It was two majors who had missed him at Quantico and flew down to get ten posters signed.

"The Marine Corps museums at Quantico, San Diego, and Parris Island love him because of the money they make on his signature. The poster sells for five or six dollars, but if it's signed, it sells for ten dollars."

"I don't get any money," Taliano interjects. "The extra is considered a contribution. So when I sell 500 posters for ten bucks apiece for one of the museums, they're making $3,500. If you look on eBay, you see posters from 1978 selling unsigned for $152. They've used different paper over the years. Selling them on eBay is not fair. What I do now is stipulate that posters I sign not be sold commercially.

"We've never truly identified the recruit in the photo. We were in the squad bay, during recruit platoon pickup. It was the very first day they were organized into their platoon. I was assigned to that platoon only for that activity, so I don't know the platoon, I don't know that recruit, and I'm sure I was just having a conversation with him, pointing out that he was eyeballing the area or needed to stand up straight, something like that. We had twelve drill instructors who were picking up a platoon, and three of them were going to be their permanent drill instructors. There were eighty-six recruits, so we just kind of got in their face. But, three years later, I had no idea who he was when the poster came out. The Marine Corps Association [MCA] has tried to find him over the years so they could do a reunion. But 1968 was a very difficult year for the Corps in Southeast Asia. I'm married to that man for life, and I don't want to know that he's dead, wounded, or maimed.

"In March of 2003 I was at Camp Lejeune signing posters, and a young gunnery sergeant told me that that recruit was his company gunnery sergeant several years back. He said the gunny had that poster behind his desk, and you could not mistake him because the eyelashes were long, just like in the photo. And I showed him another picture I have where the face isn't cropped and he said, 'Yep, that's him.' He said the man's name was Pitts. We're trying to find him. If he is alive and and a retired sergeant major, I'm sure the MCA would be delighted because that could be a whole new recruiting campaign."

Why has the recruit never stepped forward to identify himself?

"Well, if somebody was eating your ass, you wouldn't want to come forward either.

"Some things don't change. I do think there is some difference between recruit training now and in our day. Lou and I used to pick up kids who were criminals. You know, the judge would tell them, 'You can go to jail or you can go join the Marine Corps.' Also recruiters in those days, in order to meet their quotas, would sign on overweight kids or skinny bodies that needed strengthening. It's a different world now with different young men—and women— coming through boot camp. Many are in much better shape than they were in our day and, because it's an all-volunteer corps, the standards are much higher. They're better educated and smarter, too.

"At the end of the day, I feel there was much less emphasis on some of the physical acitivities and the discipline. Now people would debate that with me, I'm sure, particularly the drill instructors today. The approach today is different, but by the end of the day, it comes out to being pretty much the same thing.

"The D.I.s today face the same pressures we did during Vietnam, and I see it on their faces when they come into the museum or around the base. You can tell by the stress around the eyes. You can tell they're having the same feelings Lou and I and others had, wanting to give the recruits all they need to survive in combat. It takes a special breed to do this. That's why you gotta love the Marine Corps. You gotta love it to no end. You just love what you're doing. When those recruits graduate and when they step off, when you tell them, 'Dismissed!' the feeling that you have is much like fatherhood. And yet you're not far from them in age."

Marine Corps Recruit Depot
Training Schedule (Men)

PARRIS ISLAND

PHASE ONE

While recruit training is generally considered to last twelve weeks, it actually consists of thirteen weeks, with the first week given over to processing and preparing to learn how to be a recruit. Training is divided roughly into three phases, and is presented here as such. It is explained by Staff Sgt. Kristopher Wylie, a senior drill instructor, who enlisted in April 1999. In August of 2005, at the age of twenty-nine, he had been on the drill field three years and a senior D.I. for one year. His tour was to end in March of 2006.

WEEK ONE

Local recruits who come anywhere within an eight-hour radius of the base arrive in fifteen-passenger vans. In the case of Parris Island, those from farther away are flown into Savannah Airport and transported in charter buses. Recruits are brought in after 11:00 P.M.—in

Their Mothers Never Told Them: Uneasy recruits, just off the bus and a few moments on the Yellow Footprints, make their way into Receiving, where they will undergo Forming, or processing, prior to the start of recruit training at Parris Island, South Carolina. *(Photograph by Larry Smith)*

Parris Island or San Diego—and are immediately told in no uncertain terms by a very firm drill instructor to disembark at once and form up on "my" Yellow Footprints painted on the street outside the Receiving Barracks. They are then taken into Receiving and placed in platoons which will train as units over the next twelve weeks. Their heads are shaved and they receive the "bucket issue," consisting of clothing and various articles of hygiene. They are told to call home to say they have arrived safely. Then they are granted the "Moment of Truth," in which they are expected to reveal anything inappropriate, such as drug use, they may have concealed from their recruiter. Some admissions may be overlooked; some may be cause for an immediate trip home. The first three to five days are spent in Receiving. There is no sleep the first day: Night becomes morning as they undergo medical and dental checkups, turn in their civilian clothes, which are stored in a warehouse, and receive field gear, a pack, canteens, poncho liners, boots, and running shoes, called Go-Fasters. They are issued M-16 rifles, which will be constant companions. The rifles will even accompany the recruits to the chow hall, although they are left outside.

They are taken to the squad bay, assigned racks to sleep in, and shown how to make them up. They undergo additional administrative work and are given service record books and identification cards to prepare for Saturday, when they will meet their drill instructors. Also on Saturday, the fifth day, they take the IST, the initial strength test. They must be able to perform two pull-ups, forty-four crunches, and run a mile and a half in 13 minutes and 30 seconds. Women recruits must be able to perform a 12-second flexed-arm hang, forty-four crunches, and run a mile and a half in 15 minutes. (Those who fail are placed in the PCP, or Physical Conditioning Platoon, until they can pass the test.)

After the test, they change from PT (Physical Training) shorts and tee-shirts into green utilities and meet the company staff and their drill instructors, who give them the drill instructor's pledge. The senior drill instructor takes over and begins their training with three days of Forming. Recruits undergo additional administrative work in which the D.I.s learn "personal issues," such as which recruits are left-handed, which are active duty, and which are reservists. They make sure all gear is complete, serviceable, and ready so training can begin Tuesday. The second day of Forming, Sunday, includes church service from 7:00 to 11:00 A.M., for every denomination, ranging from Protestant and Catholic to Buddhist and Islam. Chow follows. Recruits continue with Forming Day 2. All administrative work is completed on Monday, Forming Day 3. Throughout training, recruits will receive an hour of free time at the end of each day. While there is no more telephone or physical interaction with anyone from off the base or Depot, recruits are encouraged to write home regularly.

WEEK TWO

Training Days 1–5

This is the first week of actual training. It consists of academic instruction, physical training, and the beginning of martial arts, called McMap, which is an acronym for the Marine Corps Martial Arts Program. For the first three weeks, in addition to classwork, recruits will focus on fighting with pugil sticks, so-called Tan Belt remediation, which consists of a variety of martial arts techniques involving throws, choke holds, and hand-to-hand fighting without weaponry. This is practiced daily for the first twenty training days. Recruits are tested for martial arts on Training Day 22, at the start of Phase Two. The pugil stick is a pole roughly three feet long with something like a boxing glove secured to each end. Wearing helmets and pads, recruits face each other in pairs and try to knock each other down. It is the first time in the lives of many that they have been hit by someone else. Staff Sgt. Wylie said, "It is a big confidence builder and a wake-up call to the young kids." Pugil stick fighting is based on the technique of hand-to-hand fighting with rifle and bayonet. Recruits "fight" with pugil sticks once a week, or three times during McMap, and again at the Crucible. In addition to McMap, recruits start learning drill, they do runs, push-ups, crunches, and other exercises designed to build them up physically.

Classwork focuses on the history of the Marine Corps, from its founding in 1775 in Tun Tavern in Philadelphia to the present, and on customs and courtesies of the Corps and basic first aid. Recruits learn the Marine hymn from drill instructors on their own time.

Training Days 6–11

As they proceed with various aspects of McMap and classwork, recruits also are introduced to the Obstacle Course and the Confidence Course. This is what you see in the movies. For the first time, recruits attempt the Obstacle Course, which they will negotiate three times during the training cycle and once more at the Crucible. It consists of 15 to 25 obstacles spaced about 25 yards apart over a 400-yard course laid out in a horseshoe or a figure 8, and recruits generally "execute" it in under three minutes. There are jumping obstacles such as ditches, maze-type dodging obstacles, ropes, cargo nets, or walls to climb. There are also pipes, beams, or vaulting-type obstacles. It is "one of the more strenuous things" the recruit will do "in a short time frame," Staff Sgt. Wylie said.

The Confidence Course is much more intricate, with much larger-scale obstacles, such as the Slide for Life, the Sky Scraper, the Belly Buster, the Dirty Name, and the Tough One. The Confidence Course consists of higher and more difficult obstacles, intended to inspire confidence in recruits in their mental and physical ability and to "cultivate their spirit of daring." It is designed to teach recruits to climb heights without being afraid. There are fifteen obstacles that challenge the recruit in different ways. In the Slide for Life, the recruit climbs a tower, grasps a rope, and swings his legs up and then slides down over water. In the Tarzan, the recruit walks successively high logs until he comes to a horizontal ladder, and then he grabs the rungs, reaches out, and arm-walks the length of it. Recruits meet the Confidence Course on Saturday of the second week of training. They will do it again in Phase Three and also at the Crucible.

Training Days 12–17

The recruits' first field activity is the gas chamber, which comes fairly early in order to give them something to look forward to early on in training, to let them know it's not just physical training and academic instruction. They get to go to the gas chamber on Tuesday, and are subjected to CS gas, similar to tear gas. They undergo two hours of classroom instruction on how to wear the mask, get it checked, and wear it inside for three minutes. Recruits perform four different exercises while inside.

Training Days 18–23

Initial Drill comes on Tuesday, Day 19, when platoons in the company begin to learn to march in competition with one another. A recruit is designated as the Guide for each platoon. He or she carries a flag indicating the platoon number. Later on in training, at the rifle range, the recruits design their own flags for the platoons. Training Day 19 concludes with initial drill.

CHAPTER TWO

IRON MIKE MERVOSH

Machinegun Squad Leader
Battle of Iwo Jima

There'll never be a war like that,
never a battle like that again, no way.
There were so many unselfish and unrelenting
acts of bravery, courage, and heroism that occurred
routinely on a daily basis as a true and keen sense
of duty that they weren't even recognized.

Iron Mike Mervosh stands in front of the photo of
himself made when he was a Fleet Marine Force
Pacific Sergeant Major. It hangs in the Iron Mike
Room of the staff NCO Club at Camp Pendleton, California. When this photo was taken in September of
2005, Mervosh and several hundred Marine veterans
were at Pendleton for an Iwo Jima anniversary
reunion of the Fourth Marine Division, which was
deactivated at the end of World War II. *(Photograph by
Larry Smith)*

A Marine's Marine, Mike Mervosh served two tours as a drill instructor at Parris Island. He also saw combat in three wars, holding every enlisted rank from private to sergeant major over a thirty-five-year career. He was a Fourth Division middleweight champion until wounds suffered on Saipan and Iwo Jima ended his boxing career. He holds Navy Commendation Medals for action on Iwo Jima, in Korea, and in Vietnam; a Bronze Star for action in Korea; and three Purple Hearts. He served more than nineteen and a half years as a sergeant major in battalion, regiment, brigade, station, base, and division. His last assignment before retiring, on September 1, 1977, was Fleet Marine Force Pacific Sergeant Major, the largest field command in the Marine Corps—over 80,000 Marines.

"I was born in June of 1923 in Pittsburgh, and I was a Depression boy from age six to eighteen, for twelve years, you know what I mean? Selling papers and pot scrapers and scrounging around for a buck and a half a week—anything I could get as a teenager. A lot of guys were quitting school to help their families get through the Depression, and after Pearl Harbor I went down to the recruiting office next day and the Marine recruiter couldn't handle all the applicants. He said, 'Let me have your phone number and we'll call you.' And I said, 'Well, we don't have a phone number because we don't have a phone.' He said, 'How about your neighbor?' And I said, 'None of the neighbors have phones either. We're all in the same shape.'

"This went on for months and finally my father heard about it and he said, 'What are you trying to do?' I said, 'I want to join the Marine Corps.' He said, 'You only got a couple months to finish high school.' I said, 'I can finish when I come back.' I talked him into it, but then my mother talked me out of it, see. She pleaded with me. A high school

out. I graduated in June of 1942 and I was helping out at home, huck-
stering, doing odd jobs, delivering papers, getting about ten bucks a
week. My father was laid off, all that stuff. I finally said, 'I can't be
doing this all the time. I'll go in the service, I don't know what they're
paying, but I'll send money home.' Then I found out it was thirty dollars a month, a buck a day. I said, 'By damn, that's three meals a day
and a sack and meanwhile I can send money home.' So I enlisted in
September 1942.

"We went by train to a place called Yemassee. You never forget
Yemassee. It was a rude awakening. We're not even in the damn boot
camp and these handlers, they're not drill instructors, they just handle
the troops coming in, shift them into trucks and send them on to Parris Island. And it was, 'All right, let's move! move! move!' I thought I
was moving pretty fast but that's when I got my first swift kick in the
rear end because I didn't move fast enough. It was daytime and it was
hot. It could have been anybody but I happened to be there and he
gave me a swift kick in the ass that got everybody's attention. 'Line up!
Line up! Line up!' And here's a big old cattle truck come to get us, like
you'd herd a bunch of cattle in, and we were stacked in there back to
back, hanging on to the side. I was in the middle and we went twenty-
six miles in that cattle wagon, bouncing all over the road. And then,
when we got to boot camp, we had another rude awakening.

"The trucks pulled up and everybody jumped out. There was no
ladder and we didn't have the Yellow Footprints at that time. You
jumped and if you didn't move again you got another swift kick. They
lined us up and the drill instructor says, 'From now on, I'm your
father! And this junior drill instructor is your mother! You can screw
your mother but you're not going to screw me!' Man, I'll tell you, I
thought, What the hell is going on here? Sergeant Fisk was our senior.
He was tougher than a seabag full of hand grenades. And the junior,
Private Peters, was tough, meaner and nastier than Sergeant Fisk. I
guess he was working hard for that Pfc. rank, what we referred to as
Praying for Corporal. We only had the two of them for D.I.s, and we
didn't call 'em Hats. They didn't wear Smokey hats, campaign hats,
either, just pisscutters and barracks caps. They looked immaculate,
though. Peters was terrible, I'll tell you, but I don't want to go through

all the stuff we endured in that place. I was always scared, we were all scared, they scared the hell out of us.

"The D.I.s tried to break us down mentally. We'd get a swift kick in the ass but that wasn't nothing. It didn't hurt. It was humiliation. You felt stupid. I been called stupid too. D.I.s in other platoons would hit guys, but Fisk and Peters never did. They'd yell your ear off, kick you in the rear end, and they carried swagger sticks, used them as measuring devices between bunks and once in a while they'd give you a crack on the knuckles if you didn't hold the rifle properly.

"Boot camp was eight weeks. My strongest memory was that first day. That stays with you. Our initial close order drill was performed in the sand. We had two sections, about eighty recruits. They lined us up by height, big six footers all the way down to short guys. Section One was Shitheads. Section Two was Feather Merchants. I was in the Feather Merchants. I was five feet nine and I weighed 160. What running we did was mostly for punishment. If you screwed up on the drill, then high port went the rifle and around and around you ran. We didn't have boots, we had boondockers. I remember that sand, you got sand fleas, mosquitoes, gnats, every day hot as hell. And the sand fleas are biting the shit out of you and if you raise a hand to slap one of them, boy, you've had the weenie. Then you might get 100 yards of duckwalking or other horrible stuff. Once we became fairly proficient with our drilling in sand, we were rewarded by being elevated to the grinder, the parade deck. It was asphalt and just beautiful after drilling in sand. Man, we were happier than shit when we went on the grinder. But if just one guy screwed up, then everybody went back to the sand. It was mass punishment. Later we'd grab hold of that son of a gun who fouled up, and pound on him.

"We didn't do real long marches, we did six sometimes or maybe twelve miles. We got a lot of history, special instructions in traditions, who the Commandant is, the chain of command, nomenclature and functions of your weapon, stuff like that.

"They had a bugle, played by a private, that tells the whole battalion what to do, right? First thing in the morning, he blows a bugle, get up, fall out, chow call, assembly, mail call, we knew all the bugle calls.

up, fall out, chow call, assembly, mail call, we knew all the bugle calls. After we were down there a month, there comes this strange bugle call, all us recruits looking at each other, what are we gonna do? And a D.I. comes in there, sort of pissed off, acting like we were supposed to know and, by damn it, it's pay call! No wonder we didn't know. We had our I.D., and our dog tags, and a lieutenant is there, giving us money. Wow. Not checks like you have today; I got paid in cash, for a long time, in the Marine Corps. Here I got my pay, a twenty-dollar bill, and it's the first twenty I ever owned in my life. I looked at that dag-gone thing and I said, 'Man! Three meals a day and a sack, I'm gonna stay in the Marine Corps.' Twenty bucks! I sent the whole thing home. I didn't need it. What the heck did I need it for? I sent it, no money order, just cash, to my mother. She was very gracious when she wrote back, said it was a big help.

"We had an obstacle course but it was nothing like we have now. There was a lot of marching, a lot of bayonet drill, no pugil sticks. We had Springfield '03 rifles with bayonets and this colonel who was instructing us on them said, 'This is where you kill, or be killed. But I want you to do the killing. One of you recruits, come up here and fix bayonet.' He come up to this kid and said, 'Take that scabbard off right now, and come after me. Kill me! I mean it! Go for my throat!'

"We were there a couple weeks by now and we knew something about it all, but he showed us how much we didn't know. See, that kid charged him with the bayonet and he disarmed him like nothing. He was barehanded and he parried him, and flopped him on his ass. I wish I could show you how it was done. But everybody learned how to take away an enemy rifle. We kept the scabbard on so nobody got hurt. We had dummies. You'd parry, slash and hit this way, hit 'em that way, use a butt stroke like an uppercut, and then finish 'em off.

"Did I do it in combat? My first engagement was in the Marshall Islands and I was in a trench line there and I ran into a goddamn Jap, he was about twenty feet away, from here to that wall, and he was at fixed bayonet and I was at fixed bayonet. And I thought to myself, he doesn't have ammo in that rifle, so I took my time and shot him right between the eyes. Then I started to think about it, and I went and rounds still in the clip. He was expecting me to fight him with the bayonet. He could have shot me, but I was dishonorable." Did Mervosh feel bad about that? "No, hell, no. But in his code, I was dishonorable.

"When I left boot camp I went to Camp Lejeune—it was New River then—and then we got the M-1 rifle. We trained there maybe two months and traveled by troop train across the country to Camp Pendleton. I was still a private. The training at Pendleton was very intense, very intense. We made so many amphibious landings on those beaches in California. We'd hit the beach, cross Highway 101, hold up autos, then go tactical again, just like advancing in position.

"A lot of times we'd go down to San Diego, get aboard ship in full combat gear, sail a couple days and then hit, not only California coastal beaches but also San Clemente Island. We made a lot of landings there, of course, and we killed a lot of goats with live fire. They'd get in the way and you'd kill 'em. We shot up the place—artillery, rifle fire, everything like in a real landing. Where Camp Pendleton's golf course is today, we shot up that whole place, heck yeah, with tracers and everything from .30-caliber to .81-mm mortars.

"For the landings, we'd go by boat team assignments, two rifle squads and a machinegun squad in the landing craft, you understand? As soon as you landed, you'd get the hell off the beach because you got other waves coming in, and if enemy is there, you destroy him, seize and control your objective. We'd have briefings on how far we were supposed to go, we had maps, every one, right down to the squad leader, the fire team leader, they knew what the hell to do.

"We trained for many months at Pendleton, we made so many amphibious landings, at San Clemente and the coastal beaches, combat loaded all the time. Man, I think they trained us so much just to get us pissed off. There was nothing more to learn after two or three times. You trained in misery; after a while you'd take pride in training in misery because you had to endure the hardship and rigors of battle. That's the purpose of it, get us pissed off.

"And here we go again combat loaded, down to San Diego, we're out at sea for about four, five days, all at once we got the word, this is it! Yay! Guys were glad. It was like going to college: Finally we were going to apply ourselves. They gave us the word, booklets, maps, Marshall Islands, Kwajalein Atoll, we never heard of it. Our assignment was the two little islands of Roi and Namur. They had airfields, defensive positions. If we take them, we extend the Pacific Ocean out that much farther, almost 3,000 miles from the states, a beautiful harbor, where the Navy could come in.

"I was a machinegunner. We had the light .30-calibers only. We didn't get the heavies till we got to Saipan and Tinian and Iwo. There were six of us, the gunner, assistant gunner, squad leader, and two ammo humpers. We trained so any one of us could take over if the next man fell. Even the last ammo humper knew the gun; he had to. Guys get killed. After Roi-Namur we went to Saipan and secured it in twenty-five days. We regrouped, got replacements, then had our first hot meal.

"Fire on Saipan was sporadic when we first came in, but not even an hour later they were hitting us with heavy artillery. See, they waited for the reserves to get in and the big equipment, artillery and stuff like that, then the shit started coming in, pounding beaches and boats. Guys were dying all the time. My battalion commander and my first sergeant were killed in the first hour or two.

"Then we hit Tinian, which was only two and half miles away, and took it in nine days. I was in the first wave. Tinian was the most complete and perfect landing in the history of the Marine Corps. Lot of people don't know that. We had two beaches, White Beach and I think Green Beach; White Beach was 135 yards wide, Green was 65. The Second and Fourth Marine Divisions assaulted simultaneously. But we met very light resistance because they had their master troops where the beautiful beaches were, they figured, Hell, we'd be landing there. That was what made our landing perfect because we hit where they didn't expect it. They had 9,000 Japs on Tinian alone because that was a valuable piece of real estate. Our B-29s could reach and bomb Japan from there.

"We landed in columns. The beach wasn't wide enough and I remember the amtrac I was in came up right against coral rock. There was no room for us to get on the beach. We had to jump over the sides and the water was up to our chests and you had to hold that damn rifle over your head and of course your ammo's getting wet. I was a machinegun squad leader by now and I had to jump over the side with a machinegun and a carbine. I had a squad of twelve guys, and my gunner got hit right there, got a bullet hole right in his chest. He flopped with that machinegun and the assistant gunner grabbed it. I grabbed hold of the gunner and edged him up on the coral rocks, cut my fingers on that sharp coral but I propped him up. He was gasping and all I could do was put my finger in the hole in his chest. The corpsman got there and said, 'He's gone.' And I said, 'Put him in the LVT [Landing Vehicle Tank].'

"We got to our objective but that goddamn coral—not only me but a lot of other Marines—I saw yellow pus and everything from the infection. I thought I was going to lose my fingers. Coral poison, they called it. The corpsman put stuff on it. I was scared I was going to lose some fingers. Tinian lasted nine days. They worked on me when I got back to the ship.

"We went back to Maui to rest camp after that but that was a misnomer. As soon as we got our wounded squared away, our replacements were coming in, and we were training all over again. Train, train, train, you know. We trained the dog-ass out of ourselves, again in misery, all through September and October and then came the big one—Iwo. We didn't know nothing, didn't know where we're going.

"Then they got the big layout, here's this big thing, Iwo Jima, Sulphur Island, pork chop–shaped, Mount Suribachi, and estimated how many enemy were there, the amount of live stuff, we're going to get a lot of prep fire and everything. The Navy was supposed to give us ten days' prep fire, naval gunfire, air support and all that stuff, but they cut it down to three. And all they did, all those 16-inch guns firing, the 12-inchers, 10-inchers, all they did, they just blew away some sand.

"But we had Domenick Tutalo, he was a flamethrower and he'd go up to those pillboxes and we'd give supporting fire and someone would throw a satchel charge to blow the thing and then Tutalo comes up and flames it and anyone left alive comes running out burning from head to toe, you know what I mean? So that's when we'd put 'em out of their misery with the bayonet, get some practice. We didn't shoot them, because ammo was hard to get. Why did we hate the Japs? Because they were savage and because of what they did to our country and because of the way they handled the goddamn civilians in China. On Guadalcanal they slit Marines open and cut their heads off and on Iwo Jima, when they captured a Marine, they killed him and cut his pecker off and put it in his mouth. That's how come we hated 'em.

"The enemy were dug in four or five tiers below the surface of the earth, and there was no way you could touch 'em. They had their sickbays, mess halls—everything. The tunnels were already there, dug by Korean slave labor for sulphur mining. They sold it all over the world, for use in gunpowder and making matches. The whole island is only two by four miles but the interlocking tunnels and caves, converted for military use, ran for miles. They had 642 pillboxes. The enemy were directed not to inflltrate; they were going to win this battle by attrition. Their general said no one surrenders, no one leaves the island dead until you kill at least ten marines. More than 20,000 Japanese died. They were still coming out of holes ten years after the island was secured.

"February 19 was D day, and there was never a safe place the whole campaign. We were always crouching under artillery, mortar, and rocket fire, and even their anti-tank and anti-aircraft weapons were fired at us ground Marines. They threw everything. You take the Normandy invasion, twenty-four hours after landing on the beach, you could put a picnic basket out there with your grandmother or your wife, and sit down to eat. But on Iwo, you couldn't do that after three weeks. There was no safe place. Remember when I said I got that twenty-dollar bill in boot camp? Well, later on we got up to forty-five dollars a month. It meant I was sending more money home. And I was fighting, I figured out on Iwo, I was getting $1.54 a day based on twenty-four hours, getting shot at, blown up, and so on. We were Charlie Company, First Battalion, 24th Marines.

"I had shrapnel in the leg and a corpsman told me I might be evacuated but, I said, 'Doc,' we called 'em Doc, you know, I said, 'I'm up in the high spot in Hill 362 and as I look at the goddamn beach, all the LCVPs [Landing Craft, Vehicle, Personnel] taking the wounded out back aboard ship are gettin' blown out of the water.' I says, 'I'm not gonna get blown out of that water being wounded. If I'm gonna die,

I'm gonna die here.' I didn't have to do any long walking. Hell, I stayed in the same foxhole, I remember, one time for six days, the same foxhole. Funny thing I don't remember is, I don't remember taking a crap on Iwo. Maybe we shit in our pants. Course they gave us this ration C, which was a chocolate bar, and if you had dentures you could forget about eating the damn thing; you had to use a Ka-bar to break it loose. Maybe that's what you'd call asshole cement. But that was a meal, you know what I mean? It was supposed to give you energy. I hadda take a crap, I had to, you know what I mean? But I don't remember.

"I do remember you pissed in your helmet and threw it over the side of the foxhole. You weren't going to get up to take a leak. I got very minor wounds. I had shrapnel in the back of my head, I still got a piece back there. And you can probably see a scar when I shut my eye. If I had my eye open, I'd have lost it, but at that time I caught a piece when my eye was shut. It just caught me when I blinked. I got a little scar here. That didn't cut it for a Purple Heart. I could have had five or six of them if you counted, just like John Kerry, see I'd have had a dozen like him. And a lot of Marines had that volcanic ash embedded in their skin. It was volcanic ash. Now we had a lot of casualties on Iwo, as you know, almost 7,000 Marines killed, over 18,000 wounded. Twenty-three thousand enemy dead. Now just imagine if we didn't have that volcanic ash and it was hard ground instead: The ash absorbed that shelling and saved a lot of Marines.

"I remember I was in a big shell hole getting hit hard by mortar shells. We were attacking and six of us got in this big hole because of the mortars falling all around us, and one of my troops says, 'I'm gettin' the hell out of here.' I said, 'Where in the hell are you gonna go? No matter where you go, you're gonna get hit.' I said, 'Stay here,' and, bing, a mortar hit right on top of the parapet and everyone became a casualty there except me. I opened my eyes, I mean, honest to goodness, I heard angels singin'. That sounds weird. I thought I was dead.

"I opened my eyes and I see this big gash and all this blood coming down this guy's back and everything, and he's laying there and I grabbed his first aid kit, threw some sulfa powder on top it and jammed that first-aid packet right up against his wound there and tied it around his neck. Of course, he was unconscious. We got some supporting fire which maybe counteracted that mortar fire and that's when I grabbed hold of the guy, put him on my shoulders and carried him back a couple hundred yards, and gave him to the corpsman, and they got him back in the rear. But I thought he was gone. Then I see him about thirty years later down in Baltimore; I hadn't seen him since Iwo. He said he just remembered it faintly because he was unconscious. First thing I turned him around because I wanted to look at his head, and of course he don't have any hair there, just a big old scar and I said, 'That looks pretty good. Crying out loud, I thought you was a goner.' He says, 'I thought I was a goner too.'

"You don't think too much about survival. You have a job to do, you know what I mean? I didn't have fear because fear to me is complete loneliness. In other words, there's no hope, but I felt there was hope because man, oh, man, I still got my Marines around me. It was, hey, let's go, move out, that's my job to move out, and I'm the first one to move out, I'm leading them, that's what a leader is for, that's why they called me Combat Crazy. If I die, I told the guy next to me, I said, if I go, you're taking charge, you know, and that's the way it goes because fire team leaders became squad leaders, squad leaders became platoon sergeants, platoon sergeants became company commanders and stuff like that, because of casualties. And that happened on Iwo. It was complete decimation. I always felt that, if I lost a leg or an arm on Iwo Jima, I'd be coming out ahead. No kidding. I was one of thirty-one in our company to walk off the island at the end of the battle.

"We wanted to live, but we weren't afraid to die. Does that make sense? The thing is get that enemy to die, not you. You got to have that adrenaline. We weren't afraid to die, but we wanted to live. Just like one of the Marines told me, he was in the hole there with me, he said, 'Mike, I don't know if I'll make it, probably not.' Then I said, 'Hey, we all got to think about making it; we're not all going to die,' He said, 'Well I don't give a damn if I die—I'm gonna die for my country, and one thing I know my family is gonna be safe. They're not gonna be speaking Japanese, or Jap.' We called 'em Japs. We had all kind of names for them: gooks, slant eyes, slope heads, you name it, swear words, especially when they'd say, 'Maline, you die!' 'Yeah,' we'd say, 'You son of a bitch, you're going to die first, though.'

"There'll never be a war like that, never a battle like that again, no way. There were so many unselfish and unrelenting acts of bravery, courage, and heroism that occurred routinely on a daily basis as a true and keen sense of duty that they weren't even recognized. They were taken for granted and uncountable. By today's standards I would say every Marine on that island would rate a Silver Star. Vietnam is when things changed, damn it. I'll tell you, it irks me, what they give awards for. Here's a guy wearing three rows of ribbons and he's never been shot at, never even smelled gunpowder, and he gets these Mickey Mouse awards. You take the Navy Commendation Medal, which was a really good combat award in our days: By damn, you had to clean out bunkers and pillboxes and kill people to get something like that. Not today. I believe in an awards system, but you got to have standards.

"I hate to say this, but, you know, the survivors who came back, I said, Well, we got to get ready for Japan, I said we got to train these replacements and get ready to do it all over again. The guys are calling me Combat Crazy again. 'Look,' I said, 'I'm not combat crazy, I'm combat orientated. We're fighting for our country, damn it, our families, we know they're safe, because we took the Japanese homeland, Iwo Jima. Every island we took was Japanese mandated. We didn't recapture anything. We took their islands. Our flags flew on all those places.'

"The flag-raising on Iwo? I got a story on that: You know they got the Iron Mike room in Camp Pendleton, display cases and stuff. I couldn't give 'em everything I had because they didn't have enough room, but I put these binoculars I carried on Iwo in there. I had them around my neck when we landed and I noticed I was getting shot at more than the troops, and that's when I put the binocs in my pack. Then I wasn't drawing so much fire. Well, they looked for guys with map cases, binoculars, radios, they were prime targets. Those binoculars stayed in my pack until D plus 5, when the flag went up on Suribachi.

"They had a smaller flag up there earlier, but within hours they changed it. Some colonel wanted it for a souvenir. They got the other one, the big one, from an LST [Landing Ship, Tank], and up top they found a big piece of pipe. Joe Rosenthal of the Associated Press was hiking up there all this time, and he got the picture of the flag-raising. He had to have split-second timing because it wasn't posed or anything. He caught that damn thing going up, and now it's history. He said, I liked what he said, very modestly, he said, 'Anyone could have taken the picture. The Marines took the island.'

"We didn't get the word at the Quarry, where I was at, and I didn't hear no sirens or whistles. I was up there saving my ass and about three hours afterward we got the rumor the flag was up. Now I could see Suribachi but I couldn't see the flag, so that's when I pulled out the binoculars to take a look and, sure enough, I see it and then bing! bing! bing! The Japs are shooting at me! One even caught me in the cartridge belt. My exhilaration was not from seeing the flag go up. My exhilaration was, 'You sons of bitches are poor shots.' But them frigging binoculars went back in the case, back in the daggone pack and that was the last time I used them things.

"But one time the binoculars and their case and my C-rations did keep me from getting killed. I was in this terrible terrain up north, hunched over in a crevice to get away from mortar fire and something hit me in the back, like a baseball bat, oh, like a mule kicking me, wham, you know? And when I had a chance to look, all my C-rations were gone, my binocular case was gouged and my pack was torn up by a chunk of shrapnel the size of a pancake. It would have gone right through my back. But of course that's normal.

"We had a company of 240 when we started and, by the end of the battle, 31 of us were left. Half the company had been wounded, the others killed. Coming off the island was complete exhaustion, complete exhaustion. Tired, weak, I don't know how much weight I lost. We all lost weight. Anyway, the thirty-one of us we were relieved, and we walked back to the beach, got on the by damn LCVP, pulled up to the ship and some of the Marines were so weak we could hardly climb up the cargo net. The sailors were helping some of the Marines and I was struggling, I didn't have no strength, and a sailor come up to try to help me, I said, 'Get away! I don't want no Swabbie to help me! I'll make it. I made it this far, all I got is a couple more feet.'

"Aboard ship, I didn't want to eat; all I wanted to do was crap out. I didn't take a bath or anything. They just let us go. Heavy sleep, I mean heavy sleep, and then we shaved, and showered, oh, my gosh, I changed my underwear. I had changed my underwear one time all the time we were in battle, only changed the undershirt. I would never take my pants off. My socks I got changed one time. I took my shoes off. I did that once. We must have stunk: Can you imagine the smell? My feet were all right, though.

"The first meal we had was sauerkraut and weiners and, man, I gorged myself. I barely got out of that by damn mess hall and I seen a big GI can. Bleaahhhh, my stomach couldn't hold that much, it had shrunk, I threw it all up. I sat down for a while, took a couple deep breaths, got hungry again, and went back to eat. But I didn't gorge myself the second time. Sauerkraut and weiners.

"Back on Maui, everyone was glad when the A bomb went off because that meant we didn't have to go fight in Japan. I went back to San Diego on a small carrier, there was a band and we got a nice greeting and all that stuff. Then at Pendleton, the saddest part, we had a ceremony, a parade and then the Fourth Division was deactivated. I was with it from start to finish.

"Between 1945 and Korea, I was back at Camp Lejeune with the Second Division, making Mediterranean cruises, we made some amphibious landings in the Caribbean, I went up to Norfolk, Virginia, for more training. I was with the Second Division by that time.

"My brother was killed in Korea with the Seventh Marines. He was shipped out before me. I got the word when I was on a Mediterranean cruise and that's when my mother requested that I come home with his remains. They got me orders as an escort, from the Mediterranean to New York to escort his remains to Pittsburgh. Then we buried him. I went on to Lejeune and I wasn't there but a couple weeks when I got orders for Korea.

"I wanted to go. That's what we were living for. That's what we get paid for. Crying out loud. One year and one day in Korea. I was on the DMZ, the demilitarized zone. It was mostly trench war, land mines up the gump stump. We had outposts for our MLR, main line of resistance; we had Outpost Reno, Outpost Vegas, and Outpost Carson City, and the Chinese overran Reno and Vegas and that's when we had to make a company assault and retake them. They were key positions because once they get those then they start moving, you don't know if they're going down to Seoul again or not. The peace talks were going on but that's bullshit. So we had to recapture those positions. What a hairy night. They had complete disregard for life. They'd attack by masses. They didn't give a damn, just masses, bugles, and everything. We'd have a field day mowing them down. I never seen so many by damn Chinese in all my life.

"I was company gunnery sergeant, taking care of all the enlisted troops. The company commander commands them but I tell them what the hell to do, down to the nitty gritty. I made sure I had every casualty listed, I didn't care if he was killed or wounded. I didn't want any MIAs. I didn't want that. I kept tabs, had my little black book. When we attacked them we'd get hit by artillery and I saved a couple lives, pulled them away, and the First Sergeant spotted me doing that and that's when I got the Bronze Star. I said I don't want the fricking Bronze Star but that's when I got it, for that and for rallying the troops. Well, you're supposed to rally the troops.

"We were just one company with a half-mile line to defend. My area had three outposts. It was called the Jamestown Line. We were there eighty-four days. It was so cold, I had to pee on my hands to keep them warm and I had to pee on my rifle to make it operate. We had M-1s.

"I went to Parris Island as a drill instructor in 1952, right after Korea, until 1955. They were getting tough then about how to treat recruits. The colonel on that first tour said, 'You can't call 'em turds. Look, we don't call recruits turds. We call 'em privates and boots, that's all right, but we don't call 'em turds.'

"I said, 'Okay, Colonel, I won't call 'em turds.' So I start thinking, spell it backward, you know, I said, 'All right, you druts!'—that's turds spelled backwards—and the colonel he approved of that, he didn't get it. I called 'em druts. But the recruits knew. That was before Ribbon Creek. Training hadn't changed all that much since I was a boot. You didn't even see an officer. And we didn't graduate our pla-toons. They ask me now how many platoons did I graduate and I say none. We didn't go for this picture taking and stuff. You didn't get the Globe and Anchor until the last day and then we'd call 'em a Marine—maybe.

"But I had my recruits in the swamps or at Ribbon Creek many times. They'd screw up or something and I'd march them up there at night. McKeon would have got away with it if he hadn't been drinking. That's what killed him: under the influence. If I'd have lost those men, I'd have had a court-martial too. But all of us got away with it then.

"Then I went back to the Second Division, returning to the drill field from 1957 to 1959, not long after Ribbon Creek. They brought guys in from all over country. We were supposed to clean up the place and all that. I thought we'd get a welcome aboard from the general: You're experienced drill instructors, glad to have you, welcome aboard, and so on. Instead it was, Here's what I want you to do, you follow the Standing Operating Procedure word for word and you people are going to see that it's done. And this is what I want done: You won't even blow your breath on these recruits. I mean he went on and on for about an hour. No drinking: There will be no more of this, and no more of that. You will do this, you won't do that. Boy, what a welcome aboard. He laid it on the line. I wish I had a tape of it. His name was General Wallace Greene; he later became Commandant.

"That second time at Parris Island I became a chief drill instructor, and I had charge of twenty D.I.s or more. I'm going to make sure they do it like you're supposed to. Training was ten weeks at that time. It went back up to twelve later on.

"Why does Parris Island have such an impact on young people joining up? It's the first time he's away from Mom and Dad and maybe he comes off the street someplace and he's kind of loose. He watches TV, he smokes, he runs around, and suddenly he finds himself in this place where it's harder than anything he's ever done in his life. But he gets through and when he comes out, all of a sudden he's squared away, you know what I mean? Orders I got in the Marine Corps instilled me with pride and leadership, dedication, devotion to duty, and all that stuff. Ours is the toughest recruit training in the world. The kid's got a new sense of himself, he's got purpose, he's got focus. He thinks he's ready to fight but he's not; he's still got a long way to go.

"My tour lasted two and a half years and then I went to Okinawa, Third Battalion, Ninth Marines for fifteen months, right up to the first part of 1960.

"I went to the Seventh Marines when I got to Vietnam. I was a pretty senior sergeant major by now and I could have taken division sergeant major or even regimental, but I said I wanted an infantry battalion. I wanted the Third Battalion, Seventh Marines, because my brother had been in it, although I didn't tell anybody that. My colonel came from the ranks and he and I got along great.

"It was a squad leader's war and we went for a lot of walks in the sun. They told me I didn't have to go but I said I'm going out there with the troops, I want to see what the fuck this is all about. When I die I want to die with my boots on. I went on every damn operation we made. Then I finally got up to regimental sergeant major. I got the helicopters. We flew all over the by damn place. I got shot down twice. One time I was with Colonel Hall, he's a regimental commander now. They called him Bulldog Hall. Both of us were uglier than a seabag full of rear ends. They called us the bookends, Iron Mike and Bulldog, but the troops loved us.

"Combat in Vietnam was a lot of in and out. You fought the VC guerrilla warfare, and the night belonged to them, putting out land mines and stuff. I talked personally to every Marine who joined the regiment, from sergeant major down to private, and the main thing I told them was now you're going to go out in these infantry companies, you're going to go on these doggone patrols, and most of the time you're not going to see anything. But you got to stay alert, stay alert at all times, I don't give a damn what time of day it is, because as soon as you put your guard down, you are dead. And watch the mines, pungee sticks, booby traps. In Vietnam we had more casualties from mines than from any type of small arms. If a leg was lost in Vietnam, you know what happened: mines.

"Once I had fifty guys join the regiment, and I told them just a week ago we had a squad leader out there with twelve guys, he's been doing this for months, day in, day out. They get a little tired, they get under a nice tree, nice shade, they're trying to rest a little bit. No perimeter defense, they crank off a half-hour of sleep and one damn gook with a by damn AK47 wiped out the whole damn squad. I said you'll be doing that for months and not even see one of them guys and all you have to do is just crank off one time and you are dead.

"When the colonel and I got shot down we were just landing and the pilot got in the wrong LZ [Landing Zone], a good thousand meters away from where we were supposed to land. The colonel and I were going to jump out and take a look around when an RPG [Rocket Propelled Grenade] hit that son of a gun and kind of blew both of us out just as we were jumping. We both hit the ground head over heels. Five seconds earlier we'd have been in that chopper. Pilot and crew were all killed, and it rumbled down the hill there. I got an M-16, the colonel's got his pistol, and I says, 'We don't have a radio.' He says, 'Sergeant Major, we're not going to be POWs.' That's what I wanted to hear. He said, 'You got your rifle, I got my pistol, we'll kill as many of those bastards as we can before we go down.' 'Good enough, Colonel,' I says. We were going to move out at night because the colonel was wounded and we'd never make it in daytime, 400 meters as the crow flies but a lot of up and down. Our people knew something was wrong and I'd say in about an hour a chopper come down there and made two or three passes, shooting the area out, then he come down and I said, 'Get going, colonel!' He forgot about the wound and we jumped into that SOB and they were shooting at us. You could hear the pings. We had four or five bullet holes in the fuselage when we got back. I was in my late forties.

"I retired in 1977 and in 1991 I volunteered for the Gulf War. First I called, then sent a nice letter, had it typed up by my wife to make it formal. I didn't get any response at first but then this Colonel Huly called me from HQMC [Headquarters Marine Corps]—he's a three-star general now—and he said, 'Sergeant Major, we got your letter and some of the officers told me about your call.' He said, 'We'd be glad to have you back on active duty in the Marine Corps. We checked your records and you have a fine combat record but we got to give our youngsters a chance. I'll tell you what we can do: We can have you relieve the sergeant major at Camp Pendleton, take his job and we'll get his butt over there.'

"I said, 'No, you keep him at Pendleton. I want to be over in the Mideast. I'm not volunteering to go back to being base sergeant major. I had that duty before.' He said, 'Well, we looked at your date of birth.' By now I was in my late sixties. I said, 'Colonel, you're discriminatin'.' He laughed. I said, 'I'm physically fit, I have combat experience and I can save lives. And beside you got a lot of sand in that by damn place and I'm a professional at loading sand bags.' He laughed again and about a week later here I got something from HQMC—hot damn, orders! That's what I'm thinking. My wife said, 'You always said you'd like to die with your boots on before you die of any disease.' I said, 'Yup, I'd rather go down that way. This is it, honey.' But then I opened the letter and it said, 'Sorry, our regrets. We understand your patriotism and dedication and all you did thirty-five years in the Corps.' It was a nice, beautiful letter, but . . . sorry.

"I always had my own way of doing things. Like back when I was a drill instructor: One tour, I had a bunch of these damn bandits from Philadelphia. They came down almost like a platoon. We squared them away eventually but, cripes, first we had to have a shakedown: knives, brass knuckles, every damn thing. I said, 'Well, this is a challenge. We're going to see how tough these guys are.' I said, 'You think you're gonna be tough? I'll show you how tough it's gonna be.' They made it, though. They had to."

CAMP PENDLETON

Trainees at the Marine Corps Recruit Depot in San Diego learn to shoot and undergo additional training, including the Crucible, at Camp Pendleton, which is the West Coast counterpart of North Carolina's Camp Lejeune. Pendleton is less than fifty miles north of the Recruit Depot, which is not large enough to accommodate gunfire and large-scale maneuvering.

Pendleton consists of more than 125,000 acres of varied terrain, with more than 17 miles of shoreline bordering the Pacific. When it was acquired in March of 1942, World War II was under way and the Corps needed space for division-sized exercises. Originally known as Rancho Santa Margarita y Las Flores, the site was part of a Mexican land grant dating from 1841. The Corps paid about $4 million for it. President Roosevelt came for the dedication on September 25, 1942, and the base was named for Major General Joseph H. Pendleton, who had long pushed for a base of its size. Women Marine reservists arrived in 1943 to help run the base and perform some of the clerical work, "freeing men to fight."

In 1946, the Marine Corps Commandant, General Alexander A. Vandegrift, decreed that Pendleton would be the center of all West Coast activities for the Corps. The base was home to the First Marine Division. Quonset huts were built, barracks were renovated for officers' quarters, and the commissary opened in 1948. During the Korean War, a combat town was built to simulate a North Korean village, although the Marines had to travel across the state to the High Sierras for cold-weather training. More than 200,000 Marines passed through on their way to the Far East.

By the time of Vietnam, the Korean village had become a jungle village with booby traps designed to simulate conditions in Vietnam. Marines arriving at Pendleton received fifteen days of intensive training before going on to Vietnam.

Hollywood films shot there included *Sands of Iwo Jima, Flying Leathernecks, Battle Cry,* and *Heartbreak Ridge.* Pendleton remains home to the 1st Marine Expeditionary Force and two subordinate commands.

It has firing ranges for everything from pistols to 155-mm artillery, landing beaches, parachute drop zones, bombing and strafing ranges, three towns for the practice of urban warfare, and, of course, areas suitable for large-force maneuvers. The first recruits to fire weapons there arrived in August 1964. The range is named for Major General Merrit A. (Red Mike) Edson, who received the Medal of Honor for action in the Solomons in September 1942. He led the Marine battalion that became known as Edson's Raiders.

CHAPTER THREE

BILL PAXTON

Sergeant Major
1953–1983

I'd lost my dad at Iwo Jima,
you see, so I didn't have a father
and I lost my my uncle. That's why I
wanted to become a Marine. My mom said,
"Are you crazy? After what happened to your dad
and your uncle?" I said, "That's why I
want to be a Marine."

Former Sergeant Major Bill Paxton continued to work at the Marine Corps Recruit Depot in San Diego following his retirement in July 1983. Paxton, who was a leader in the campaign to create *The Known Marine* memorial to Marine Corps drill instructors, was seventy when this photo was taken at the base in September 2005. (*Photograph by Jean Fujisaki*)

If you call Bill Paxton on the phone (in September of 2005), and he is not available, this is what you hear: "Oorah! You have just reached Sgt. Major Bill Paxton, USMC Retired, but still active. At the sound of the tone, please leave your name, the date and time of your message. May we never forget our fallen comrades! Freedom isn't free. Once a Marine, always a Marine. Semper fi. Oorah!" I first heard him telling stories and counting cadence at a D.I. reunion in Parris Island, and we finally got together outside of San Diego, where he lives today in Santee.

"I was born July 11, 1935. I grew up in Indianapolis, Indiana, and graduated from an old Army arsenal that had been converted to a high school, in June of 1953. My recruiter was Lt. Archie Van Winkle, a Medal of Honor recipient from the Korean War. He was a staff sergeant who got a meritorious commission in combat. I was going to quit high school and join the Marine Corps with my future brother-in-law, but when Lt. Van Winkle got to checking my papers, he said, 'You got good scores but you don't have a high school diploma; you just got a half-semester to go. What's wrong with you?' I told him I wanted to join with my future brother-in-law, and he said, 'Bullshit. Go back and finish school, and then if you still want to be a Marine, come and see me. Better yet, I'll be at your graduation.' I thought he was bullshitting me, but he was the first to congratulate me at my graduation, even before my mom.

"Because I'd lost my dad at Iwo Jima, you see, so I didn't have a father and I lost my uncle. That's why I wanted to become a Marine. My mom said, 'Are you crazy? After what happened to your dad and your uncle?' I said, 'That's why I want to be a Marine.'

"And Archie said, 'Do you still want to be Marine?' I said, 'Yes, sir.' He said, 'Well, you got the Ernie Pyle Company leaving in one week. Meet us there at the Federal Building.' I'd already taken the test and everything. Ernie Pyle was a famous war correspondent that got killed in the Pacific near the end of the war. There were about 160 of us, three platoons; we made up a whole company.

"We went to boot camp in San Diego, come out here on a troop train. I'll never forget one recruit who got in the wrong sleeper car. He had the women all shook up. Then he took off for a beer in the lounge and that was where the drill instructor caught up with him and he had him going back and forth through our sleeper car all night long, saying, 'Sir, I am a shitbird; my heart pumps shit, not blood.' We pulled in here at the Marine Corps Recruit Depot in San Diego and the train backed up into the Depot right near where I park my car today, every morning when I come to work, right by those same railroad tracks where I got off the train for boot camp. They had Bob Hope and Miss San Diego, and they were singing, "Back Home in Indiana," and then they played the Marines' Hymn. As soon as Hope and his crew and Miss San Diego pulled away, all hell broke loose. That was when we got introduced to the duck walk.

"More than thirty years later, after I retired, I went to work at the Recruit Depot in San Diego. We issue all uniforms to recruits and to students at the drill instructor school and recruiter school. We rotate from year to year in different parts of the job, working in alterations one year and maybe in shipping and receiving the next. I've been here nineteen years as of September 2005. It's a paid civil-service job. The USMC forever—that's my billet. I'm retired but still active.

"Boot camp then was sixteen weeks, counting processing. It was tough, it was damn tough. They still had the real bucket issue: You got a bucket for scrubbing and cleaning, for keeping all your gear in, and you sat on it to clean your M-1. It served a lot of purposes. I won't talk about how they put the bucket over your head to mark time with a swagger stick. I was a platoon guide but I broke my hand and went to the hospital and they kept me there for rehab and put me in a working party, so when I went back, I had to start all over again. So I was in boot camp almost six months. And the rest of the company all got dress blues and they got to come home to Indianapolis for a big parade, and marched around the Monument Circle there, where they'd sworn us in.

"But I got out of boot camp in late November of '53 and the company was long gone. I finally met up with them when they came back out for infantry training at Camp Pendleton. I got a leave after that, took the train to Indiana. When I came back from leave they stopped the train in Pasadena on New Year's Day because the Rose Bowl Parade was crossing the tracks. I remember I looked through the train window and saw Roy Rogers and Andy Devine on their horses, part of the parade.

"I went to sea school in San Diego, learning marine nomenclature and the history of the seagoing Marines. I was O3, infantry, but we found out later sea duty was a special assignment. We had to go before an interview board at boot camp before we got selected. And then, while we were waiting orders to go aboard ship, they assigned us to a silent drill platoon. We did all the same drill that they do at "8th & I," the Marine headquarters in Washington, D.C.

"We performed all over California for almost six months and then I finally went on the USS *Kearsarge*, CVA 33, Carrier Vessel Attack. I was the captain's orderly, and captain of the ship's bodyguard. We had the special weapons post, we had the brig, and we had the ship's landing party.

"We went overseas, to Japan, crossed the equator, and when we came back, they made up a new detachment and we recommissioned the USS *Oriskany*, an aircraft carrier that had been converted to an angled flight deck. It supported submarines and the rest of the fleet, destroyers and cruisers. It was a challenge for me because I wanted to be in the infantry—that was what I signed up for. I was a grunt."

Paxton came back to San Diego and served in the military police, just missing duty with the State Department because his divorce was not finalized and Marines had to be single to serve embassy duty. He got sent to Camp Matthews at Pendleton as a rifle and pistol instructor, and then volunteered for a tour in Hawaii.

He was to serve a second tour in Hawaii, 1962–1965, when he was reunited, eight years after enlisting, with his recruiter, the Medal of Honor recipient Archie Van Winkle, who was by now a colonel in charge of the Escape and Evasion School at Kaneohe Bay. "I was with the Fourth Marines," Paxton recalled, "and I was the sergeant in charge of the aggressor detail at the Escape and Evasion School. We would harass and capture everyone who went through the school, even the officers, the helicopter pilots who thought they didn't need it, but they found out they did because once their choppers went down, they became POWs. We'd capture them and take all their chow, tie them to a banana tree, put molasses on them, and let the ants give them a little indoctrination.

"Colonel Van Winkle was just a Marine's Marine. He would call us out at night and we would go steal the weapons [of the men attending the school] while they were asleep. Next morning, he'd say, 'I want that rifle number,' and the lieutenant would go, 'Uh, sir, I can't find my rifle.' And Archie would say, 'You can't find your rifle?' Then he'd say, 'Paxton, bring the lieutenant's rifle up.' And he would tell the lieutenant, 'In combat you'd have been dead.' He really laid heavy on those pilots.

"I came back to San Diego in '64 and went on the drill field for fourteen months. It was during this time a series gunny came by and said two recruits were slipping off after taps and making illegal head calls. He was sure they were smoking in the head, but the drill instructors couldn't catch them. They would flush the butts down the commode and take off out the back hatch. I said, 'Gunny, none of my recruits would do that.' He said, 'Well, keep your eyes out anyway.'

"So I got my two junior drill instructors, Sergeant Hoy, who was later killed in Vietnam, and Sergeant Zug, and I told Sergeant Hoy to go in the head and get up in the rafters, and I told Zug to go in the other end of head and put a bar across the back hatch, so if there is anyone in there, they won't be able to get out. I said, 'I'll come in the front hatch and we'll trap them. I'll tell you goodnight, see you in the morning, and then you double-time around behind and get in the head.' He said, 'No problem.' So anyway Taps went, and I said goodnight. Well, they took off and did their thing and it wasn't five minutes before I hear the pitter patter of shower shoes going to the head. The two recruits come out the back tent, and we run over to the head and we hear, 'Hurry up! Hurry up, man! Light up! Light Up!' I came in, put the bar across the hatch, and flicked the light switch. Sergeant Hoy jumped down and grabbed the one just as he lit the cigarette and caught him before he could flush it away. One recruit says, 'Sir, sir. It wasn't me, sir. It was him, sir. I just had to make a head call.'

" 'Okay,' I said, 'you sit on that commode and you're not going to get off until I see a turd.' What hurt was they were in my platoon. I told Sergeant Zug, 'Put them in a dipsy dumpster with a lit cigar and let it get good and smoky for about ten minutes, then take and put them in the shower and then put them in the rack.' I said to Sergeant Hoy, 'You take the platoon to the classroom after chow, and Sergeant Zug, you make sure to get those two out of the dumpster.' Fine, no problem. Well, morning came and I went to check on recruits at dental, check on boots, eyeglasses. Then I met Sergeant Hoy in the classroom and said, 'Where's Sergant Zug?' He told me Zug got called to the duty hut right after I left, was told his wife had miscarried and he went on immediate emergency leave.

"I said, 'What about the two recruits in the dumpster?' He said, 'Well, you told Sergeant Zug to get them out.' Sure enough, we did a head count and we were missing two recruits. I took off out to the dumpster and the dumpster's gone. I go to the Back Forty, where they empty all the trash. They had the dumpster up in the air, it's smoking, the fire department's hosing it down, and I said, 'Oh, no.' So I said to the fire chief, a former Marine drill instructor, 'Oh, no, what happened?' He said, 'Staff Sergeant Paxton, no problem, no problem. I saved their ashes, whatever you want to do with them, throw them away, military funeral, whatever you want to do with them. No one knows but you and me.' Then he said, 'But don't ever put me in this position again.'

"I said, 'Oh, no, oh, no.'

"Then he said, 'No, no. They're both over there under that palm tree. We got a poncho and a blanket around them. They're good to go.'

"I took 'em back. I said, 'You two just passed the most rigorous top-secret mission in the Marine Corps. Not a word to nobody, if you want to see your parents again, you understand me?' They said, 'Yes, sir.' Oorah.

"Anyway, I kept them both as squad leaders, even put them in boxing smokers. I found out that one had been a Golden Gloves champion before he came into the Marine Corps.

"Now this is '64, right? Fast-forward to 1982, a year before I retired: I was the base sergeant major up at Camp Pendleton. General Robinson, the base commander, said, 'Sergeant Major, I'd like you to go down to San Diego to the El Cortez Hotel. They're having a boxing fund-raiser for disabled vets. I've got to go back to Washington, D.C., and I want you to go in my place, representing the Marine Corps and Camp Pendleton, because our division commander can't go either. You can sit in the VIP row.'

"I had boxed in smokers so I said I'd be glad to go, and I took my son with me. He was in the enlisted delayed entry program at the time. We're sitting in the front row and they call Ken Norton, who had been Heavyweight Boxing Champion of the World, into the ring. They say, 'Mr. Norton, sir, will you please give us a few words?' So he climbs in the ring and takes the mike and he says, 'A few words.' And he walked off. They said, 'Oh, no, sir.' He said, 'A few words; that's all you get.' They said, 'Sir, please, how about your retirement and whatever became of your trainer?'

"So he said, 'Once I retired from the ring, an elderly lady ran an exit up there at Highway 5 in Hollywood, totaled my car, and totaled my body. My son helped me rehabilitate. He's captain of the UCLA football team. And as for my trainer, the only reason I'm here tonight is because my real trainer was my Marine Corps drill instructor. Sgt. Major Paxton, sir, will you please come into the ring?'

"I'm looking around. My son said, 'Dad, he wants you in the ring.' I said, 'Bullshit, I wasn't his drill instructor.' He said again, 'Sir, will you please come into the ring?' So I come climbing through the ropes, in full uniform. I said, 'Mr. Norton, are you sure I was one of your drill instructors?' He said, 'Yes sir, you caught me and my buddy in the head one time making an illegal head call. My buddy was smoking, but you saved our careers. Then you put me in a boxing smoker and I went on to become the Heavyweight Champion of the World—all because you give me a break when I was a recruit.'

"Then he took me all around the ring, held my hands up, and they rang the bell fifteen times. That was 1982. I retired in '83 and started working here in January of 1986.

"Well, come March of '88 and I'm checking recruits, they're signing for their cammies and their boots, and here's this Kenneth Norton Jr. I said, 'Private, who is your father?' He said, 'Sir, Kenneth Norton Sr. My father was a former Heavyweight Champion of the World.' I said, 'Oh, you stand by over here.' Then I called his drill instructor outside and told him I wanted to use that recruit Private Norton in my working party.' He said, 'You got him, Sergeant Major. Keep him as long as you want.'

"So then I had him field day [clean] the warehouse with a tooth-brush. When he finished, I said, 'Now you write your dad and tell him that his drill instructor had you field day the warehouse with a tooth-brush.' He said, 'Aye, aye, sir.' Then the week before his graduation, I got word his father wanted me to attend his graduation. I said I would be there. Graduation day, I found the father and his son, way up in the bleachers. The second son, the one who was team captain at UCLA, went on to play for Dallas and the '49ers, number 51, in the NFL. Anyway, I called them down out of the bleachers.

"He comes hobbling down, because that accident had shortened one of his legs and busted up his rib cage. I took him and his son to the reviewing stand and introduced him to the sergeant major and the general, who invited him to sit next to him in the reviewing stand. Norton said, 'Yes, sir, but I'd like my drill instructor to sit with me.' The general said, 'Well, who was your drill instructor?'

"He said, 'Sir, Sergeant Major Paxton was one of my drill instructors.' The general said, 'Sergeant Major, why didn't you tell me?'

"I didn't really answer, because I had feared Norton might refer back to that incident with the dumpster. After all those years, we didn't need that to come up. But he was loyal. If you seen him right today, if you talked to him right this second, he'd say, 'What dumpster?'

"That fire chief taught me a lesson, all right. But if anyone asked me, I'd tell 'em that's just a sea story. A sea story only. No truth to it. Oorah.

"My second tour as a drill instructor ran from September 1966 to November 1969. I was on the drill field for a while and then in Sep-tember of 1967 became a close combat instructor, teaching hand-to-hand and bayonet fighting for the entire recruit regiment.

"We put on this demonstration for General Wilson, the Comman-dant, and the commanding general of Parris Island, the commanding general of the Depot at San Diego, the mayor of San Diego, VIPs and congressmen. We showed all the different guard positions, and when the gunny announced the importance of the element of surprise, we had a Marine camouflaged up as a Vietcong in a spider hole in front of where the Commandant was sitting. When the gunny said 'element of surprise,' Sergeant Foster come out of that spider hole chattering Vietnamese and the gunny jumped down with a diagonal slash and put the bayonet right between his rib cage, more or less, and, when he rolled over, we had rubber guts with the red dye squirting out. We said, 'Corpsman up!' and they carried him off. And the Commandant said, 'As of this second, close combat is adopted.' See, we went back to the fighter's guard position so we could parry left and parry right with the rifle and bayonet and also be able to kick with our combinations.

"This was for recruit training. We were training the whole recruit series, teaching rifle and bayonet movements and chokeholds, throws, the blocks, punches and kicks, a type of karate. We combined rifle and bayonet with karate and that was why they called it close combat. What they teach today in McMap, Marine Corps Martial Arts, is an outgrowth of that. They referred to us as the dirty dozen because we had twelve instructors. I did that from September of '67 to November of '69.

"Recruit training in the Vietnam era shrank to eight weeks, which was insufficient. I had come back from my first trip to Vietnam in July of '66, then went again in 1969 and '70, when it was a lot worse than before. The first time we were there helping them out. My church sent clothes and food and toys and we would take them out to the villages while we were on combat patrols. We'd help the children, teach them how to use the soap and say thank you and you're welcome. We didn't throw stuff to them like dogs. We would line them up and show them respect, made sure the clothes fit, divided up food, and showed them how to open cans. And they would show us where weapons were hidden and where the booby traps were. They saved quite a few Marines. And they loved the corpsmen because they'd treat them well. A lot of them had been hit by booby traps too.

"We did a lot for them people that the media never covers, the country never hears about. I'm sure the Marines are doing the same thing over in Iraq, helping out those kids, trying to win their hearts over, letting them know they're there to help them.

"We saw quite a bit of action on my second tour. Sometimes it was hot and sometimes you might go two or three weeks without nothing. They did that on purpose, hoping you'd drop your guard and then they would hit you when you least expected it. You had to be on your toes at all times. We made a booby trap lane in Vietnam to show new guys what to look for. They could use our mortars but we could not use theirs. They were shrewd, never underestimate the enemy.

"If you don't get discipline in boot camp, it's too late to get it in combat. It's technique and discipline. The more we sweat in peacetime, the less we bleed in war. No Marine ever died in his own sweat. I used to emphasize No. 1, Make sure you got all your men; make sure they got all their gear; make sure they know the mission in case they have to take over. Keep the gas in the back; know where the grenades are at; the smoke, the red, white, yellow, green, CS gas, frag grenades, illumination grenades. Two green, two red, one yellow, two gas grenades for tunnels, two frags per man, all for routine combat patrol. I'd rather have too much than not enough. The average patrol was for eight hours. It had a sniper attached, plus an engineer to clear mines or booby traps. A reinforced patrol consisted of twelve with engineer and sniper and a corpsman. The size depended on the mission. The radioman was key. In one action, my machinegunner got hit, and I ran out to help him and a round creased my cheek and went into his neck and heart.

"My helicopter got shot down in an action called Operation Double Eagle. We were heading back aboard ship. We were the last team to leave that zone and we got shot down. The pilot was killed, the co-pilot was wounded, and we had to destroy the helicopter. We made a body bag out of a poncho so we could carry the skipper. We got him and the co-pilot to a Special Forces camp. They didn't have no security. We could have wiped out their whole camp. We flagged a bird down and they landed us back on the ship. I'm trying to think it was the *Valley Forge*. But anyway the first sergeant met us there on the flight deck and he was so happy. He thought he would have to list us MIA. He gave us a big steak dinner.

"Later on we were down at LZ Ross near the Queson mountains, where we had a lot of problems with mutineers at the end of '69, beginning of '70. They were Marines, part of 2-7, mostly HQ, you know, like Headquarters and Supply, where they had the cooks and then the sick, lame, and lazy from the other companies that refused to go out in the bush. They took over the compound, I'm guessing sixty to eighty of them; I think they were associated with the Black Panthers. They'd refused to go out in the bush and we had to retake our own battalion compound. We had to tie them up and fly them back to Da Nang and put them in the brig. A lot of them were court-martialed, flown back to the states, and we never did hear the outcome. I had similar problems as the regimental guard chief in Okinawa, between October and December of 1970. We had to recapture a barracks that had been taken over. No one would ever believe it if you wrote a book about it, and they would say I was prejudiced. But those that were there, they knew, and it was bad news."

In a twenty-two-page article published in the *Armed Forces Journal* (June 7, 1971), Col. Robert D. Heinl Jr. wrote:

> The morale, discipline and battleworthiness of the U.S. Armed forces are, with a few salient exceptions, lower and worse than at any time in this century and possibly in the history of the United States.
>
> By every conceivable indicator, our army that now remains in Vietnam is in a state approaching collapse, with individual units avoiding or having refused combat, murdering their officers and non commissioned officers, drug-ridden and dispirited where not near mutinous. . . .
>
> Intolerably clobbered and buffeted from without and within by social turbulence, pandemic drug addiction, race war, sedition, civilian scapegoatise, draftee recalcitrance and malevolence, barracks theft and common crime, unsupported in their travail by the general government. . . . Distrusted, disliked and often reviled by the public, the uniformed services today are places of agony for the loyal, silent professionals who doggedly hang on and try to keep the ship afloat.
>
> Historical precedents do not exist for some of the services' problems, such as desertion, mutiny, unpopularity, seditious attacks,

and racial troubles. Others, such as drugs, pose difficulties that are wholly NEW. . . . By several orders of magnitude, the Army seems to be in worse trouble. But the Navy has serious and unprecedented problems, while the Air Force, on the surface at least still clear of the quicksands in which the Army is sinking, is itself facing disquieting difficulties.

Only the Marines—who have made news this year by their hard line against indiscipline and general permissiveness—seem, with their expected staunchness and tough tradition, to be weathering the storm.

"I came back from there in December of 1970 and went to Twenty-nine Palms, hoping to get back on the drill field, but instead they made me a career planner and then I got sent back down to San Diego to take part in the first recruiter school they established here. The school used to be only at Parris Island but they changed it in 1972. From there I went full circle, back to Bloomington, Indiana, where I'd started from, this time as a recruiter. And whenever I got to feeling bad or discouraged or felt the pressure coming on too strong, I'd remember Archie Van Winkle saying, Keep the faith, keep on going.

"And I had some of the same experiences he did: I got to send some people back. I said, 'Hey, you go back and finish school.' I'd tell them what happened to me, and sure enough they'd go back and graduate. I was a recruiter from 1972 to 1976. I even made recruiter of the year." Paxton had remarried along the way and by now he had two sons, Paul and Charles, and a daughter, Laura, who was born in 1976.

"From Indiana I went to Okinawa, where I picked up first sergeant. I had a communications and electronics company. August of '77 I was back in San Diego where the sergeant major, known as Hand Grenade Wade because he lost a hand in Korea, found me a billet as the acting Support Battalion sergeant major. I made sergeant major September 1, 1978. I was squadron sergeant major for about a year and then took over headquarters base at Camp Pendleton. I was the Marine Corps base sergeant major from 1982 to '83, when I retired.

"I had my appointment and relief ceremony at Camp Pendleton but I had my retirement parade at the Recruit Depot at San Diego on July 8, 1983, the same day my son Paul was graduating from boot camp."

The Known Marine

Bill Paxton remembers:

"Gunny Sergeant Robert C. Roper was the one who gave me the idea for *The Known Marine* memorial monument. He was just a Marine's Marine and, I've said this many times, his cadence would make a dead Marine strut. He just had that rhythm. That's the purpose of cadence, to instill discipline, motivation, enthusiasm, dedication—all this comes from drill. Everything about him and what he taught us was the best, from the position of attention to parade rest, from back to attention, to the hand salute—he made sure all the fingers were extended at the joint and the forearm was at the correct forty-five degree angle. If it wasn't just right, he would correct it and tell you to glance out of the right eye, and if you can see the palm of your hand, you know you

Flanking each other in front of a marble block etched with the Drill Instructors Creed are a male and a female drill instructor. Laid out before them but not visible in the photographs is a walkway of memorial bricks into which are carved the names and Corps connections of former Marines. One brick in the front row carries the name of Gunnery Sgt. Robert C. Roper, who was killed in Vietnam. His death inspired Bill Paxton to lead the drive that led to the creation of the memorials at both San Diego and MCRD Parris Island. Paxton's brick is next to Roper's. *(Photograph by Larry Smith)*

got the salute correct. He said if you give a good salute, that shows you really care, that shows you got the pride.

"When I came back from Vietnam and was returning to the drill field and had to go through drill instructor school again he was there teaching history. He wanted me to come out first in the class because we'd served together and he wanted any Marine he had confidence in, he wanted that man to do his best. It was my own fault I didn't come out first, but he wanted me to.

"He had me do a little spiel about Vietnam because at that time only four of us in the class had been there. I had this old Mark 4 dummy grenade and when you pulled the pin and the spoon popped, the hammer would hit that primer and it would sizzle like it was going to blow up. A little smoke come out of it. I said to the class, 'You must never underestimate the enemy. You never know when you are going to be hit with that element of surprise,' and I rolled that grenade down the middle aisle of the classroom. With the smoke and the sizzling, people were going out the window, under the chairs and all that.

"Then Gunny Roper says, 'Staff Sergeant Paxton, get 'em back in the classroom.' So I hollered and motioned for them all to come back in. When he graded me, he gave me a 98, said it took too long to get them back in the classroom.

"Well, Gunny Roper and Gunny Garcia, the general's driver, volunteered for Vietnam. They were there just a short time and they got hit with an RPG, a direct hit that killed the company commander, Gunny Roper, and a radioman. After they brought the bodies back, we had to go up to San Francisco in September of 1967 to identify him. The only way we could tell it was him was he had the Marine Corps emblem tattooed on one arm with the saying 'Death Before Dishonor,' and on the other arm we could see part of a rabbit tattoo. His nickname was Rabbit because he ran all the time.

"I was one of the pallbearers and when I presented the flag that had covered the casket to his widow, Virginia, she started pulling it apart, saying, 'No one will ever hear of my gunny again.' That's when I got to thinking, You know, she's right. If he had been suspected of maltreatment of a recruit, he would have been on national TV and in the headlines, but because he completed two successful tours plus served as drill instructor plus requested mast so he could volunteer for Vietnam, all he made was the back page of the *Navy Times*: Gunnery Sgt. Robert C. Roper, USMC, KIA.

"That was the respect that the veterans of Vietnam got, even after they got killed. It was bad enough to come home alive and be treated like, you know, but to have that happen to a man who gave his life for his country, that's when I got the idea, we need to get a memorial monument for all the drill instructors.

"It would not be just an *Unknown Soldier*, because something went wrong with the Army, they got an unknown soldier. There's no such thing as an 'unknown Marine.' Marines take care of their dead. So we called our idea for a monument *The Known Marine*."

It was to take nearly thirty years. Paxton eventually went to Washington to gain approval of the plan from the Commandant, General Chapman, who told him the Corps would support it but Paxton and the others involved would have to form an association to raise money and be responsible. Paxton was sent to the Pentagon to see the colonel in charge of war memorials, and plans went forward from there.

He said the Drill Instructors Association was formed in 1986 with the help of Sergeant Major Leland Crawford, some other retired Marines, and Mike West, who became its president.

The monument at Parris Island was dedicated April 26, 1999, and the one at San Diego, with a considerable assist from General James Jones, a former Commandant, was dedicated a few months later, on

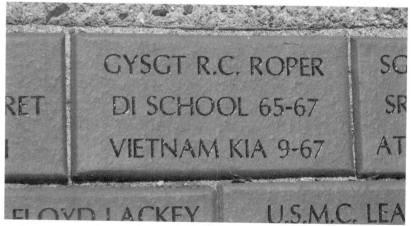

(Photograph by Larry Smith)

September 11. They cost approximately $150,000 each. Paxton said he opposed the inclusion of the statue of the woman drill instructor in the monument at San Diego, but he got outvoted.

To help raise funds for the monuments, the Drill Instructors Association came up with the idea, which continues to this day, of selling for $100 each engraved bricks, which are then laid in front of the memorials at both Parris Island and San Diego.

At the San Diego memorial, among the bricks laid in the front row is one inscribed: Gunnery Sgt. R. C. Roper, KIA 9-67, Vietnam.

Marine Corps Recruit Depot
Training Schedule (Men)

PARRIS ISLAND

PHASE TWO

Having learned how to be recruits and to understand what the drill instructor wants, recruits have now begun to get the hang of it. They have done a good deal of PT, negotiated the Obstacle and the Confidence Courses, and been through the gas chamber and learned pugil stick fighting and some hand-to-hand combat. Part of the rationale for "Phasing" is to divide the training up so the recruit finds short-term rewards to help sustain the motivation needed to complete boot camp. Now comes Phase Two.

Training Day 20

This is the beginning of Swim Qual, or swim qualification week. Wearing full utitilities, recruits will be required to swim twenty-five meters in shallow water, jump off a ten-foot tower, and swim ten meters to the side of the pool. The conclusion of the basic qualification is flotation, tread water, and using inflated blouse and trousers to help stay afloat. Everyone must meet that standard to graduate from recruit training. Those who will be in aircrew or other specialized duty must undergo higher levels of qualification. Recruits who do not pass the basic test the first time are given additional days in which to qualify. Nonswimmers are given an entire week to learn. Combat water survival instructors say they do not teach recruits how to swim; rather, they teach techniques to survive in an "aquatic environment." On Saturday, they have academic instruction and drill.

Training Day 26

Administrative activities follow swim week. Recruits have individual photos taken in a waist-high, front-fitted dress uniform that is open in back. On Friday they are given their first written exam, a fifty-question test, multiple choice, requiring 80 percent correct for a passing grade. That Saturday is the five-mile hike to the rifle range, with the ALICE

Staff Sgt. Kristopher Wylie, a senior drill instructor, watches as a recruit, Michael J. Wagner, snaps in with the M-16 in a sitting position at the firing range at Parris Island. Staff Sgt. Wylie enlisted in April 1999. In August of 2005, at the age of twenty-nine, he had been on the drill field three years and a senior D.I. for one year. His tour was to end in March of 2006. (*Photograph by Cpl. Brian Kester, Combat Correspondent, MCRD Parris Island*)

pack, about 4,400 cubic inches. ALICE is an acronym for the combat load that goes into the pack, which for the hike contains a change of utilities, extra set of boots, socks, Kevlar helmet, and the like.

Training Days 30–35

Grass Week begins at the rifle range on the following Monday. Recruits are in the grass all week, learning how to sit and adjust to positions they will execute on Firing Week. They have the sitting position, the prone position, and kneeling position on one knee. Weapons are empty. Recruits are given roughly six hours of training over two or three days with the ISMT, the Integrated Simulated Marksmanship Trainer, essentially a video game similar to the game Duck Hunt on Nintendo, from a personal computer. Called Grouping, its purpose is to insure the recruit fires a well-aimed shot each time he or she looks down the rifle sight. The recruit will also shoot the rifle in an hour-long exercise in order to verify that it is firing accurately. This is first real interaction with the weapon. Recruits stay in squad bays at the rifle range. There are five different squad bays, each of which houses six platoons. Women occupy a separate squad bay. Grass Week lasts five days. Saturday is devoted to more practice at "snapping in" with the rifle and related activities. There is a chow hall in the area as well as four different buildings available for use as chapels for religious services.

Training Days 36–41

Firing Week consists of learning to shoot the M-16 A2 service rifle from 200, 300, and 500 yards. Each recruit will shoot fifty rounds in one course of firing. Targets are maintained from pits manned by recruits who pull them up and down and mark shots of recruits on the firing line. Once a recruit fires a round, the target is pulled into the pit, marked by a recruit with a shot spotter that indicates where the bullet hit. The recruit on the firing line will plot his shot in a rifle data book, where he records all his firing throughout the week. Primary Marksmanship Instructors, or PMIs, in addition to individual coaches, staff the range and direct recruits. Each coach oversees three recruits. The PMI is responsible for the overall platoon. A great deal of emphasis is placed on safe handling of the rifle throughout.

Minimum qualification for the recruit is 190 out of 250. An Expert score ranges from 250 to 220; Sharpshooter, from 219 to 210; and Marksman, the basic qualifying score, 209 to 190. Insignia are awarded for each level. Friday is qualification day, but there is also a prequalifying day on Thursday in the event of bad weather anticipated next day.

Training Day 41

A ten-mile hike takes place on the Saturday following Qualification Week. Recruits wear the ALICE pack and carry canteens and their weapons. Roughly five stops are taken in which recruits can remove the packs, sit down, drink, and made head calls at portajohns situated along the route. A lead and a following vehicle accompany the marchers for safety, as well as two Navy corpsmen. Because all of Parris Island is only twenty-six miles in circumference, the hikes go in big circles. This hike begins at Firing Range and concludes back at the squad bay. Noon chow follows the hike.

CHAPTER FOUR

ED WALLS

Sergeant Major
1954–1984

Did the Marine Corps change my life?
Oh yes, big time, because if I hadn't gone in,
I'd probably have been in jail somewhere by now
because I used to raise hell.

Gunnery Sgt. Ed Walls receives the Navy Commendation Medal (with Combat V) for service in Vietnam from January 22, 1970, to January 11, 1971, from Col. William F. Gately Jr., the commanding officer of the H&S Battalion at Parris Island, on March 25, 1971. Walls was the brig warden at the time, after having survived a year of intermittent combat in the area of Hill 55, not far from Danang. "Days, weeks would go by with nothing happening," Walls recalled. "We'd walk and look, walk and look, and always be watching out for booby traps. They tore us up." *(U.S. Navy photograph. Courtesy of Ed Walls)*

I met Ed Walls and his second wife, Lillian, and their dog, Murphy, "part poodle, part pomeranian," at their comfortable suburban home in Beaufort, South Carolina, only a short drive from the Parris Island base where Walls had spent so many months training recruits. They were extremely gracious. Lillian commented near the end of my visit on how she "had never heard Ed speak of these parts of his life before." I promised her a tape. It was early in 2004, twenty years of being settled in one place after all the travels Walls had made with the Corps. The home was a far cry from their beginning: "The only place she went with me during those years was up to Camp Lejeune," Walls remembered. "When I married her and went to Lejeune, I was a buck sergeant and she was staying with her mother in Chestertown, Maryland. I went up to the housing at Camp Lejeune and I said, 'You got any houses?' They said, 'Yeah, we got a trailer. You can get in right now.' So I signed up and called her and said, 'We got a trailer down here. It's over in Camp Knox.' She said all right. She had a 1950 Chevrolet—it had everything I owned in it—and she met me at the circle there at Camp Lejeune. She said, 'Where's the trailer?' So we went over there and as you go into Camp Knox you see these civilian trailers privately owned, nice long trailers, and I said, 'Damn, this is all right.' Then Lord and behold, we come to our number and it was painted silver, and it wasn't as long as from here to that damn end of this room. She started crying, made me feel bad, about that damn high, and I thought we was gettin' a palace. It had a table you had to put up after you ate. It had one bed in back and a sofa. I guess it might have been 25 foot."

"I was born in 1930 in Chestertown, Maryland. I was twenty-four when I went into the Corps in 1954, so I was the old man of the battalion. I retired August 1984, after thirty years and one month. Before joining, I was in the merchant marines from the age of seventeen, working on a tanker for the Sun Oil Company out of Marcusville, Pennsylvania. I assisted on loading the tanker and doing the steering. Went all around the world although usually our run was to Texas and back. Ever once in a while they'd charter us out to South America or to the Suez Canal or the Persian Gulf to deliver oil. I got a lot of experience at a young age; it was good money back then, too.

"I didn't like school, I hated school and I quit school after seventh grade. I thought I knew it all. All my sisters and brother went to high school but I just didn't want to go. I worked for a while at the latex plant in Dover, where they made women's girdles, stuff like that, and when I come out of there I went in the merchant marines and got married.

"After six years I come to the point where I realized I didn't actually have a trade and I was tired of going to sea, being out to sea all the time, and said, Well, I got to do something, I've got to go in the service. So I went to see a recruiter, and he asked if I was married. I told him yes and he said, 'We're not supposed to take married people so tell 'em you're single.' So I come down here to Parris Island boot camp and told 'em I wasn't married. About three-quarters of the way through boot camp, my wife at that time needed money and she wrote to the commanding officer. They called me in. I took a right good chewin' and a couple boots from the drill instructor. I remember the lieutenant colonel, he had those silver oak leaves and, my God, they looked big, because back then you didn't have any officers around recruit training, no series officers or nothing. Anyway, the only thing he asked me was, 'Are you married?' I told him, 'Yes, sir.' He said, 'Why'd you lie?' I said, 'Because I wanted to come in the Marine Corps.' And he said, 'Do you want to stay?' And I said, 'Yes, sir.' He said, 'All right, you go back to your platoon and if you don't hear from me any more, you can stay.' So I never heard from him no more, and I graduated, and I stayed.

"I got paid and they also give my first wife an allotment. It wasn't very much, but it was a lot of money back then. I got some but never saw none of it till we finished. Boot camp in '54 was right rough, mainly mentally. Physically it wasn't that bad because, to my recollection, we never had any physical training per se like on the schedule; it was always drill and ever place you went you had to run. You didn't walk. You doubletimed ever place, didn't matter where. There was no officers around and the drill instructors could pretty well do what they wanted. In fact, my junior drill instructor was Pfc. Alvarez. Back then a platoon would graduate and they would take the honor man, who made meritorious Pfc., school him a little bit, and make him a drill instructor, just like that. Alvarez was the third Hat. The senior D.I. was a Staff Sgt. Dunn, and we had another D.I., and, like I say, they did just about what they wanted with us.

"Alvarez was the meanest one, believe it or not. I don't remember him touching anybody but my God he broke all rules—he'd put a recruit in the garbage, in the dipsy dumpster, and then the damn truck would come and be getting ready to empty the dipsy dumpster and they'd get him out just before they emptied it. He'd think he was about to be dumped into the truck with garbage; Alvarez did it just to scare the daylights out of him. They've pretty much done away with the duckwalk now but then if they got mad at you, why, we'd sometimes duckwalk for two days for fumbling our training.

"We usually went to bed ten o'clock and got up at five and we lived in Quonset huts with no air conditioning or nothing. In July, when I went through boot camp, the bugs were terrible. The sand fleas would tear you up and you didn't dare touch one of them. It was terrible, but, again, that's where the mental part come in, because you didn't touch one out of self-discipline or will power. The drill instructors didn't touch them either and it was just as bad on them. You knew they were human too." Walls said the D.I.s would eat the sand fleas.

"There was swimming and the Obstacle Course and at a certain phase of training we would go out to Ellis Beach and spend a few nights playing war games and climbing underneath the barbed wire under live ammo. We don't do that today, oh, God, no. And, like I say, quitting wasn't an option. If you didn't keep up with the platoon, they'd put you back in with another platoon and you'd have to start all over again so you figured you might as well go ahead and do it and get the hell out of there. But when I look back at it, it was a lot of fun. Training—this was in '54, right after the Korean War—lasted twelve weeks. We had graduation but there wasn't any pinning or anything.

We just got right aboard the buses next morning and went up to Camp Geiger for infantry training.

"Today's recruit is more physically fit, probably, than we were but I don't feel today's recruit is as mentally tough as we were back then. That's my opinion from watching over the years and dealing with recruits. They preached that you had to make it, and that was all there was to it, where now if you really don't like the Marine Corps, hell, you can get out because the Marine Corps don't want you if you don't wanna be there. In those days, you had to make it regardless of how long it took you unless you were medically unfit. But today it's no problem to get out. I don't know if that's good or not. Of course they're smarter today than we were and in a lot better shape. But when it comes to self-discipline, I don't think they're as good. They're still good people, don't get me wrong, and they do their job, but I just don't feel they got the discipline we had back then.With the equipment they got today, with the weapons and stuff, me with no education, I proba- bly couldn't have gotten in. I was a buck sergeant, E-4, a young drill instructor, before I took the high school equivalency test. I got my GED at Parris Island, took the test in the library.

"Did the Marine Corps change my life? Oh, yes, big time, because if I hadn't gone in, I'd probably have been in jail somewhere by now because I used to raise hell. I had plenty of money when I was in the merchant marine and I partied all the time. That was part of the rea- son I come to my senses and said I've got to do something. When I went in, I knew then I was going to make it a career. There wasn't any doubt in my mind.

"My MOS was infantry and I went up to Geiger, then got my orders, and, lo and behold, they put me on sea duty for two years, and here I'd just joined because I was tired of going to sea. I went aboard an air- craft carrier, the *Ticonderoga*; we went on two cruises to the Mediter- ranean. We'd go to France, Italy, all over Europe.

"Later I became the admiral's orderly and I stayed with Admiral Harris until I got orders to Camp Lejeune. I was there two years and then reenlisted and put in my first tour in Japan, in 1958. I had got divorced while I was on sea duty and had one son from that marriage. I met Lillian, my second wife, when I came back from Japan, at the

American Legion in Chestertown, Maryland. We got married after a very few weeks and it's been forty-five years now. We got married in March 21, 1959. We had two girls and a boy.

"In 1961, we moved to Maryland when I was sent to Marine Headquarters at 8th & I in Washington, D.C., where I made staff sergeant. How I got there I don't know. I was on the drill field and all of a sudden I got these orders for 8th & I. I was up there on that Presidential Honor Guard from '62 to '65. I was slim and pretty squared away and looked good in blues and everything. I stayed three years and it was a very, very interesting tour. I stood guard over the *Mona Lisa* in the National Gallery when it came to Washington and I had the death watch on President Kennedy in the White House. Then General Douglas MacArthur died and I had the death watch on him, standing at the corner of the casket as people filed through. He was in the Capitol Rotunda one day, then we flew him to Norfolk, Virginia, for two more days of lying in the Rotunda there. It was hot and the wax on his face kept melting and the mortician would come and build it back up and then later you could see the wax melting before your eyes. It gave me nightmares.

"I left 8th & I in '65, and went to Vietnam in '66. I think it was with Charlie Company, First Battalion, Ninth Marines. We went to Danang and then went out south to Hill 55 and worked out of there. I was platoon sergeant. We probably had thirty-six guys, maybe over forty if you had a machinegun attached. My first tour was thirteen months. The second was twelve. They pulled the Ninth Marines out of Vietnam and sent us back to Okinawa to retrain and regroup because we'd lost so many people. We made a landing in the Mekong Delta and when we come out of there I stayed to the rear and got ready to come home the next week. The sergeant that relieved me got killed leading the platoon down the Street of No Joy. It could have been me. I was home on leave when I got a letter from the gunny saying Clark got killed.

"I saw quite a bit of combat. It was hard to make sense of: It was very frightening. You'd just walk and walk and walk, and you'd go into little villages. A lot of times we'd go through one, get on the other side, and they'd start firing behind us. We'd turn around and expend a million rounds but you didn't know what the hell you hit. We'd move as a platoon unless we were going through a village. Then we'd be on a skirmish line.

"We ran into a couple of ambushes but mostly what hurt us was booby traps. Very seldom would you see the enemy and get into a real firefight. A booby trap could be the dud of a 355 shell that the VC would take and rig up. Or we'd throw away batteries from a radio and they would get the batteries and make booby traps out of them. Off Hill 55, there was another hill a unit of the Army of the Republic of South Vietnam was living on. They went back to Danang and they left a bunch of bouncing betties on the damn hill and VC moved in and got all those land mines and we hit plenty of them. You step on that mine, it bounces up about five feet and explodes and takes your legs off. That's usually where it got you. They tore us up.

"We went back to Okinawa to regroup and retrain a little bit and we were back there I think for a month then we got aboard ships and they took us to the Mekong Delta and inserted us by helicopter. My company went in first to set up for the battalion, and my platoon went in to set up for the company. They put all kinds of fire and missiles and stuff in there first, then about dawn they helo-lifted us in. We set up a perimeter waiting for the rest of the company to come in and—it was amazin'—that evening we were settin' there when the tide come in and goddamn before the night was over, we were standing knee deep in water. Hell, we didn't know anything. It was coming up through these roots in the trees and we were standing in water till daybreak. Then the battalion come in and we started sweeping through different areas until they pulled us out. I think that operation lasted three weeks.

"I was a buck sergeant in 1961 on my first tour as a drill instructor. I volunteered. I loved it. It was the only job in the Marine Corps that I had that I could see what I was doing from hour to hour. I mean, from the time you picked 'em up at recruit receivement and started calling cadence for them, you could see what you were doing. It was very satisfying. You know, you go to Lejeune you play war games, and, hell, you don't get to see the big picture. But as a drill instructor, you do, and it is very satisfying.

"You had the McKeon case, Ribbon Creek, in '56, when things started to tighten up. They were told to fix it or they were going to do away with the Marine Corps; Ribbon Creek damn near sank the Marine Corps. I come on the drill field for the first time in '60. I didn't know McKeon but it was after that they more or less opened up PI [Parris Island] to the civilians. They had series officers then. There had been a lot of changes. When I went through boot camp, from what I'm told, the Standard Operating Procedure manual, the SOP, only had two or three pages and of course now it's like a book. You don't do this, you don't do that. When I first came on the field, we still did a whole lot of stuff but wasn't nowhere near as bad as when I went through. You had series officers and everybody watching you all the time.

"I was probably on sea duty when the drownings occurred. I heard about it. Oh, my God, that thing went through the Marine Corps like wildfire. I don't know what McKeon did wrong; he did no more probably than what people did ever day back then. Like I say, there were no series officers or nothing. I think when he got 'em down in that creek two or three of 'em probably panicked and that's what made 'em drown. I would say they probably had full packs because a lot of times, for punishment, you would put on a field transport pack loaded and everthing—some of 'em would load 'em with sand and dirt to make them heavy, then march 'em till they dropped. But I think it was just poor judgment. I really do.

"That's a very deep creek. You can go up there in a boat. He probably marched them on the edge of it and the running tide pulled some of them out, and that's what did it. When I was a drill instructor, we'd march them out to Ellis Beach, where there's marsh on both sides of the road. We would let the recruits get out there and we always told them when we blew the whistle to hit the deck and course we'd get 'em out in the middle of that marsh and blow that whistle and they'd hit that deck and get a little bit of mud and stuff on 'em. I did it when I was a recruit going out there. But as a D.I. I never pulled any weird stuff and I never did anything unless it was for a purpose. I'd boot 'em in the ass or punch 'em in the gut or something like that but it was for a purpose; I didn't do it just for the sake of doing it.

"You'd punch 'em or put a foot in their butt or PT [Physical Training] 'em till they couldn't PT no more. Of course now you've got certain amount of push-ups you're allowed to do or a certain amount of bends and thrusts. Well, back then we'd just do it till they couldn't do it any more, but there was always a purpose behind it.

"We had one drill instructor in the Second Battalion, the same one I was in. A recruit went in to breakfast right after he got him and—this is what I don't buy because it makes all the rest of us look bad—the boy I guess was all uptight and excited eating breakfast and of course you had to eat everything you put on your plate. Take all you want but eat all you take. The boy got sick on the tray, and he made him eat it. Now to me that's uncalled for. And of course he got busted, a court-martial for it. Made the rest of us look bad too. That's the reason the SOP kept getting bigger and bigger: We wrote it ourself.

"A lot of changes in the SOP followed McKeon, of course, and we had officers involved but still a lot of stuff went on. Even so, from hour to hour you could see improvement in kids. I've got letters to this day from people thanking me 'for making a man out of my son.' I got one from, God, he was a congressman or something, and, hell, I thought I was going to jail when I saw it was from the U.S. Congress. But he was only saying he was sorry he couldn't be at his son's graduation but, 'as you know, we've had trouble with the *Pueblo*.' You remember the *Pueblo*, that spy ship? He was over there investigating that, and I can't think of his name, but I got a letter from him and as soon as I saw it, I thought, 'Shit, that boy told on me, I'm going to jail.' But he was thanking me.

"Once you got out of boot camp and up to Lejeune things loosened up a little bit. I liked to say you can always start out tight and loosen up, but you can't start out loose and tighten up. That don't work. That's the way I always tried to operate. In Lejeune you get more freedom. You think for yourself. Over here the drill instructor thinks for you, all the time.

"I was a three-stripe buck sergeant my first time on the drill field in 1960. The tour was usually a year but I extended for a second year, 1961. My second tour as a D.I., around 1967, I was a staff sergeant. When I come back from Vietnam that time, I worked one platoon as a junior and then went to senior. They only let me do a year and then sent me right back to Vietnam in '68, for twelve more months.

"The oversight on drill instructors was getting tighter and tighter each time I went. On my second tour, I had two guys working under

me. I went in one Sunday morning—it was my time to take the duty—
and I looked up on the second bunk (you had bunks that you slept two
high), and I saw this seabag up there moving. I said, 'What the hell is
that?' And it was our little short guy we used to call him The Scribe.
He'd take care of our paperwork and stuff. He was just a little short
guy, and that damn junior drill instructor who'd had the duty before
me had put him in that seabag and hung him up on that damn top
bunk and put another seabag over his head. And here it was about
time for the series gunnery sergeant to come around because his son
delivered Sunday papers and the recruits were allowed to buy a Sunday paper then. Damn, I was scared. I said, 'Man, if somebody comes
and catches us, we've had it. That's it.' So we got him down right
quick. He didn't yell. He was just in there moving around; he'd done
something wrong, I guess. I chewed that junior out, told him don't
pull none of that stuff.

"I got selected for sergeant major when I came back from Okinawa
in 1978. I went up to New River, North Carolina, and then got orders
back to Parris Island and I took over the Women Marines Battalion. I
was sergeant major there for less than a year and went with General
Christmas at First Battalion, stayed with him awhile, then went to regiment as sergeant major there and then in '81 I went up to Depot sergeant major, went back overseas for a year, come home and retired.
My highest command was here at the Marine Corps Recruit Depot. I
worked for the commanding general. I was his right-hand man. All the
other sergeants major, enlisted people on Parris Island, were under
me. I'd go down into the battalions and talk to sergeants major and
other people and keep the general updated on how things were going.
Him and I got along very good, just like I did with General Ronald
Christmas in the battalion.

"I always tried to lead by example. I wouldn't ask anybody to do
anything I hadn't done myself. When people asked my opinion, I told
'em. They might get mad and everthing but I always tried to tell 'em
the truth. I wouldn't lie to them just to get on the good side of them or
anything like that. I was never what you'd call 'well liked' at Parris
Island because I was always right hard. I was always respected. I
didn't ask anybody to like me, as long as they respected me.

"There's only two people in the Marine Corps who wear a star and that's generals and sergeants major. You don't have a sergeant major at anything below a battalion, and you're always the right-hand man for the enlisted personnel of your commanding officer. And the more senior you get, you progress to regiment or division or depot. There's one sergeant major of the Marine Corps, and he works for the Commandant. He wears two stars in between his stripes.

"You don't have to be a college graduate to have leadership neither. And you're not born with leadership. It comes from experience. I always believed that if staff NCOs did their jobs right, they made their officers what they are. So if you get a staff NCO who leads an officer the wrong way, you will see that officer go only so high, and you'll see that officer lacks competent staff NCOs. I got a nice letter from a colonel who used to be CO of an air station over here. He wrote me, 'When you were my sergeant major I learned more from you than I did from any school I went to.' It made me feel good.

"I was with General Christmas almost a year before he was relieved at regiment. He got the Navy Cross fighting door to door in Hue City, Vietnam. He never talked much about it. I thought the world of him and Sherry, his wife. When he made lieutenant general, I thought he was going to become Commandant. Evidently he was not on the team. To get up there in that big stuff, you got to be on the team. He'd have made a good one, I'll tell you that. We were tight. He depended a lot on my advice, and I put everything into it.

"Of course, he limps from that wound he got in Hue City. His whole damn heel is blown off. Him and I used to run at lunch time, four or five miles together, and he'd go back and sit down in that office and take that shoe off, take the bandage off and, goddamn, it'd be bleeding. So when we had change of command for me and the sergeant major who was going to relieve me, we were out there practicing, and of course General Christmas was standing up there at the head because we had to march up to him, and we were almost there and I started limpin' a little bit, and he said, 'You wise ass.' Oh, God, I laughed. But you got to know the attitude of your commander.

"There is no comparison between the Marine Corps and the other services. No comparison in discipline, in looks, wearing the uniform, or anything. And it all starts right here at Parris Island. The Army, I don't knock 'em, they got a job to do, we got a job to do. But we've got the most demanding boot camp training in the world. We've had Army officers come down here, we've had British officers come over here and watch, and it all starts at Parris Island. The Army is just not as strict on things and they have women and men in the same barracks. We're separate on that. We have the Women's Recruit Training Battalion.

"As I said, I was sergeant major for less than a year with our Women Marines Battalion before I left to go with General Christmas. There's a place for women in the Marine Corps, yes. I feel they can relieve a man from a desk job to go out and carry a gun, but I do not think a woman should be in combat. I really don't. If we were to have a big war and women were in combat, they could be taken prisoner and raped and stuff like that. I do believe training should be the same because, if you're a woman and you go to war, you might be behind the lines but you don't know if those lines are going to be broken and women have to be able to protect themselves, they've got to learn how to shoot. When I was sergeant major of their battalion, them gals did the drillin' and the shootin', same as the men.

"They didn't have the Crucible but now they do that too. That's a good thing the Commandant, General Krulak, put through, little old short Krulak, Shorty. And don't you look down on him when he come to inspect, neither. He'd ask, 'What the hell are you looking at?' Don't you look down. You look straight ahead. He wasn't too tall, but don't you let him know it.

"I was ready to get out when I got out because it wasn't as much fun as it used to be. My last tour overseas in Japan, the year before I come back to retire, was good for me because I had a chance to figure out what I was going to do, which I didn't know at that time. I guess I was getting older; I'd pretty well had enough. I enjoyed every day in the Marine Corps, don't get me wrong, but after thirty years I was ready to go.

"I come out and I lucked out. I went to work down to Hilton Head as a security guard and have been doing it in one form or another ever since. Right now I'm doing security at the Technical College of the Low

Country; it's like a trade school down there. I like to stay busy. Even when I'm home on weekends I'm out in the yard working, doing something. I had a boat right up till three years ago. I used to fish quite a bit. I've seen my friends retire and the first thing you know they're settin' in a bar at eight o'clock in the morning. And that's not me."

Camp Lejeune

Like Camp Pendleton on the California coast, Camp Lejeune in North Carolina serves as a major East Coast training site. Originally called New River, Camp Lejeune consists of 246 square miles containing various bases on the Atlantic seaboard. Newly graduated recruits go there from Parris Island to obtain their Military Occupational Specialty and to undergo infantry and other training. Originally, the War Department purchased 11,000 acres in 1941 for amphibious training and other purposes. Although Pearl Harbor had yet to be attacked, World War II was already under way in Europe, and military planners were anticipating America's entry into the conflict. The ports at Wilmington and Morehead City were nearby.

On May 1, 1941, Lt. Col. William P. T. Hill was ordered by the seventeenth Commandant, Major General Thomas Holcomb, to establish and assume command of the base, then known as Marine Barracks New River, North Carolina. His original headquarters was situated at Montford Point. In August of 1942, it was moved to Building #1 at Hadnot Point. Near the end of 1942, the base took on the name Camp Lejeune, in honor of the thirteenth Commandant and commanding general of the 2d Infantry Division in World War I, Major Gen. John A. Lejeune.

Camp Lejeune and the satellite facilities at Camp Geiger, Camp Johnson, Stone Bay, and the Greater Sandy Run Training Area have a long history and play a strategic role in preparing Marines for their specific job in the fleet. More than 12,000 Marines undergo Marine Combat Training at Camp Geiger annually.

Camp Johnson, which now serves a crucial role in the follow-on training of thousands of Marines every year, was the first training base for black Marines and was originally known as Montford Point. Black Marines attended boot camp here when the Marine Corps was still racially segregated. It was subsequently named in honor of Sgt. Major Gilbert "Hashmark" Johnson.

Outside the gate of Camp Johnson stands the Beirut Memorial. Once a year the Marine Corps commemorates the death of 241 Marines who were killed on October 23, 1983, when a truck packed with explosives blew up their barracks at Beirut International airport. They were part of an 1,800-member Marine force which was itself part of a multinational "peace-keeping" force stationed in Lebanon as a deterrent in the conflict involving Israel, Syria, Lebanon, and the Palestinians.

Today, Camp Lejeune has fourteen miles of beach capable of supporting amphibious operations, fifty-four live-fire ranges, eighty-nine maneuver areas, thirty-three gun positions, twenty-five tactical landing zones, and a state-of-the-art Military Operations in Urban Terrain training site. Military forces from around the world come there for bilateral and NATO-sponsored exercises.

The base is home to an active-duty, dependent, retiree, and civilian employee population of nearly 150,000.

CHAPTER FIVE

DAVE ROBLES

Sergeant Major
1958–1988

Dave Robles was my battalion sergeant major
when I was a recruit. While my battalion commander
was giving me the Band of Brothers speech, Sergeant Major Robles
was standing off on the side, and I never heard him say a word, ever, but
I looked at him, and I said to myself, *I want to be that guy.*
—Rob Bush, Sergeant Major, Third Battalion, Parris Island, May 2005

Dave Robles in his dress uniform as a sergeant major. Robles survived a gritty childhood and later, being bounced from the Corps for fighting and carousing, before he managed to get back in and become a role model. He turned sixty-five in April of 2005. *(Photograph courtesy of Dave Robles)*

I first met Dave Robles when I went into his little store of military collectibles, Der Teufelhund, to buy a fatigue hat during my first visit to Parris Island in the spring of 2004. The place was a jumble of "stuff I wanted to get rid of that I'd accumulated over thirty years in the Corps," Robles told me. He was wearing the hat I wanted, and he had no other, so he gave that one to me. He said he had started the store ten years earlier but now, he said, he had more junk than he had when he started: Marine Corps–issue clothing from different eras, bayonets, pins, insignia, and the like. "I get a lot of business," he said. "My wife says I need to run it as a business instead of a hobby and my tax lady says I got to change also. I can't run it out of my back pocket any more. What's a Teufelhund? That's what the Germans called the Marines after facing them at Belleau Wood in World War I. It means Devil Dog."

"My folks were from a Mohawk reservation southwest of Montreal called Kanawakee, and they came to the States in the 1930s. I was born here. I'm the oldest of ten. We lived in Elizabeth, New Jersey. I ran away when I was fourteen years old. I loved my parents but it was just one of those things, spur of the moment: I was walking down to the YMCA and I had my sneakers. I was going to play basketball or soccer. It was wintertime and I was walking with a couple buddies. We were talking and I said, 'Screw this,' and I kept on walking to the highway and stuck my thumb out. I had no idea where I was going. I ended up at the bottom of New Jersey because I didn't know there were other roads than that. So then I hitched all the way back up to Elizabeth, found a major road, and hitchhiked down to Texas. I had a dollar and I bought a candy bar and a loaf of bread. Back in those days, fifty, sixty years ago, people picked you up. They fed me.

"I ended up in Houston and lived there for a couple years. I worked at a restaurant, bought a Harley Davidson motorcycle, paid the rent, and went to school down there. I had a social worker because I was a runaway, and she would check on me once a week. She got hold of my parents and my dad came down. He was in the merchant marine and his ship pulled into one of these ports and he drove up to find me and said, 'Hey, what are you gonna do, are you coming home, or what?' I was an independent little shit. I said, 'Well, when I get ready, I'll come home.' But I finally went.

"Back home I sold the bike, hung around with my buddies, then went in the Marine Corps, right to Parris Island, in July of '58. It was two years after Ribbon Creek and when I look back I can see how that influenced the Marine Corps. There was a training schedule, there was PT, everything was organized. It wasn't like before, when you could train any way you wanted to and and there were no officers around.

"But there was still a lot of thumping when I went in, and I got a thumping the first day coming off the bus. I'll never forget this: We were all from New York and New Jersey. We came down on the train to Yemassee, did the police call [helped clean up] at the train station, got on the bus to come over here. The receiving drill instructor gets on the bus, hollers something, 'Knock off the grab ass!' Well, I never heard that term grab ass before, and I said, 'Who's grabbing whose ass?' You know? And he comes back with his campaign hat and he starts swinging, hits me right on top of the head. Man, that raised a big knot. I said to myself, 'Gosh, I don't wanna be here.' But everybody shut up.

"So now, having had that happen, we get off the bus, and we're standing in the Yellow Footprints, first night we get in, we have breakfast SOS [shit on a shingle], ground beef with gravy, two pieces of dry bread toasted, we get that down, and we're all lined up in some kind of formation and the D.I. says, 'All right, run by and grab a helmet liner and put it on your head!' Well, again, I'd never seen a helmet liner, so we run by and I grab this thing and put it on backwards. I just put it on backwards. So he goes up and down the line and he goes, 'Oh, a wiseguy,' and he takes it off and—boink!—hits me right in the forehead with it. I said, Oh, my head, my poor head. First I'm hit with a campaign hat and now I get hit with this helmet liner, right? Oh, Jesus, I don't want to be here. I said this is gonna be horrible. I'm a kid from New Jersey, you know? If anybody messes with you, you punch him. But this drill instructor would kill me—and this is still the first day. I'm going, Oh, gee, I'm getting my ass beat for nothing, you know?

"So we get some bags and stuff and he says, Okay, follow me. We all about-faced and I was one of the four guys in the front and the drill instructor says, Okay, follow me. He steps off and I try to catch up with him and I step on his heel and I pull his shoe off. Uh-oh. I don't wanna be here. I'm lookin', I didn't know where I was, but I tell you, that was the first day. And it was so hot, just soaking wet, and three times I get beat.

"After that it seemed to get easier. Everybody said, don't be in the front, be in the middle. It wasn't easy. I wasn't used to that kind of regimentation, but I made up my mind, I said, I'll do it whether I like it or not, you know?

"We lived in old wooden barracks that no longer exist. You washed your clothes by hand on Friday or Saturday, depending on the schedule, and you brought them up to be inspected by the drill instructor. If he didn't think they were clean enough, and this happened to me, he would throw them under the barracks in the dirt. So now they were wet and full of sand and you'd take them back to the wash rack, which was a concrete thing, and you scrubbed them with a scrub brush and cold water and soap until they were clean and then you hung them on the clothesline. And if the drill instructor didn't feel right, he'd come down and knock the whole thing over and you'd start all over again.

"You can't even talk about that to recruits now because everything is done by the laundry. But, to me, that was mean. Here I'd busted my butt washing my clothes and the D.I. would say, No, they're not clean, and he'd throw them under the barracks. There were other guys who did the same thing, just harass you so you'd learn attention to detail. It was more important than anything else to get them skivvies clean.

"I got to be a squad leader as a recruit and we were doing push-ups with packs on and every time someone failed do a push-up, we'd all have to do ten more. This guy named John Steadman kept falling down, a big heavy guy, and I thought he was fat. I said, 'Steadman, I'll beat your ass if you don't do those push-ups.' Finally we were done. I said, 'Steadman, get your ass in the head.' Back in those days the head was big as a room. It was concrete. So Steadman gets in the shower stall, and I go in and I lock the door, and I hit Steadman as hard as I could: Bam! Steadman didn't move. Bam! Steadman didn't move. Steadman was from Ohio; all he did was pick up 500-pound bales of hay and throw them on the truck, he was all solid muscle. And I'm trying to unlock the door so I can get away from him.

"Steadman beat the shit out of me. We were fighting really bad and we were both bloody by the time the section leader got the door open and pulled us both out. Man, I was bloody all over and Steadman was bloody all over. We were just punching each other out.

"I had this split lip and we had to go off to PT so I took toilet paper and put it on my lip. Steadman had toilet paper all over him and the drill instructor goes, 'What the hell did you two do?' 'Sir, I fell over my footlocker.' Finally we had to go down and talk to this chief drill instructor. 'What happened?' 'Fell on my footlocker.' 'What happened to you?' 'I fell out of my rack.' 'Sticking with that story?' 'Yeah.' I split my whole lip all the way through on the inside. But they said if I'd gone to sickbay they'd have sewed me up and I wouldn't have had that scar. I got a scar here on my lip here to this day, a big scar.

"Boot camp lasted nine weeks then. It's standard twelve or fifteen weeks now but during Vietnam it was cut back to eight or nine weeks. It goes up and down depending on the times. I finished boot camp in September, and went up to Camp Geiger for two months of infantry training. We had a month of mess duty, then we did our regular training, went on leave, then reported back after Christmas to Camp Lejeune. I went to the Eighth Marines and spent three years with them, went to Key West, Florida, got discharged in 1962, then came back in three months later."

Robles took being an old-style Marine to extremes, and this got him bounced from the Corps. "I got in a lot of trouble. I smuggled a woman into the barracks, I was drinking on duty, I was out of uniform. It took a long time for me to grow up. I'm still not grown up sometimes, I think. But I was fighting with civilians, the police, following my own drummer. I was in a position of authority as an MP [military policeman] but when I got off duty I did my thing, drunk and disorderly and all the other stuff. I liked liquor, hard stuff. Beer was okay, but you had to drink fifty gallons of it to get someplace but take two shots of whiskey and you're gone.

"I joined the reserves when I went back home and I was working some kind of paper-shuffling job which I really didn't like and I happened to be passing a recruiter's office and I walked in and said, 'Hey, look, I was in the Marine Corps, what's the chances of me going back on active duty?' So he said, Give me your serial number, and I gave it to him. So I went to work and a couple days later my mom says you got a phone call from this guy. So I call and he says come on down, I got your orders. I don't know how he got me back in the Marine Corps. I don't know if he needed a quota or if he pulled some strings or they didn't look at my bad record. But I got back in.

"I went back to the 8th Marines and they sent me to the same company I was in before my transfer to Key West, and all the guys were corporals and sergeants and of course I was a private. We're all the same age and the corporal is telling me what to do, and I'm going, 'Screw you,' and next thing you know here I am getting in fights again. I tell you, I was a hard row to hoe.

"The Marine Corps was forgiving to me, and I'm really fortunate. I don't tell that to brag on it. I made mistakes. I'm not a perfect guy. People say, 'Well, fucking Robles was an asshole.' Yeah, he was, and I'm the first to admit it. I'm not like some of these guys nowadays and—don't get me wrong—but they're squeaky clean. There's nothing wrong with them from the day they were born until the day they die—don't do anything, you know? Don't you shit? Don't you eat? What do you do for excitement? Dull.

"I volunteered for the drill school for my first tour, in 1964, but I got busted down to corporal for maltreating a recruit, and got sent to Vietnam the fall of '65. I know, it was big punishment.

"I did two tours over there, the second in 1968–69. My first tour was with Hotel 2-4, Second Battalion, Fourth Marines, known as the Magnificent Bastards. They still have that on their logo. I was a corporal, a squad leader. We operated down in Chu Lai, doing patrols. At that time I was carrying a single-shot M-79 blooper. We didn't have enough men, so I carried it. The blooper had sights but a lot of firing was point of aim. You could throw a hand grenade twenty or thirty feet but this goes out a couple hundred yards. We carried whole bags of .22-mm grenades for it."

Robles saw a good deal of combat. He suffered his first wound while on patrol. "I got shrapnel in my ass the day after Christmas. They patched that up and I kept going until I threw back a Vietcong grenade during a night ambush on February 3, 1966. I was the team leader, and I set out a claymore and had everybody lay down parallel to the trail and said when that goes off, I want everybody to fire one full magazine. Well, some guys came along, the mine went off, we opened fire, and one of them turds threw a grenade. There was moonlight and rifle flashes so I could see it and I grabbed it and threw it back. I got a concussion when it went off, and one of our guys got a slight wound from it.

"I lost my hearing on my right side from that grenade and two months later, in April, I medevacked out, went by ship to the Philippines, then Guam, and from there to California, then came to the naval hospital in Beaufort. My family was here in South Carolina at the time. I got a bunch of decorations, a Bronze Star for that firefight, for surviving and getting the men out. I got the Navy Commendation, Purple Heart, Combat Action Ribbon, a bunch of other stuff, I got it all that first tour.

"My second tour was with Echo 2-4, same battalion, different company. I was a platoon sergeant by then. I saw combat both times, not as much as other guys but enough. I mean, here's guys trying to kill you and they don't even know you. It was scary. They're trying to kill my ass. We were up in the DMZ the second time and it was like two different wars. I'd been gone a year and half. By that time the Marine Corps was up on the DMZ fighting North Vietnamese Army regulars where down south we had been fighting the VC, the Vietcong guerrillas. Those NVA guys were trained; they knew what they were doing and they didn't give up, or stop. They kept on coming, just like Marines, basically. If their job was to take the hill, they took the hill.

"I broke my ankle in a mortar attack and got sent home again, back to the naval hospital in Beaufort."

Being injured didn't stop Robles from having some fun. "I was in the hospital in Beaufort with a guy named Don Leveque, who'd been blinded by shrapnel. His eyes were all bandaged. One night we got to talking and he pushed my wheelchair out of the hospital with me giving the directions. We went a couple blocks down the street to this place called Joe's Spaghetti House, for wine and pasta. Two guys helped carry my wheelchair up into the restaurant. The Shore Patrol missed us at bedcheck and came and rounded us up and followed us back. Leveque later had two glass eyes made with the Eagle, Globe and Anchor embedded, and he supposedly liked to slip them into a guy's beer when he wasn't looking.

"But you know, I hear all these guys say that people were opposed to the war in Vietnam, protesters and such, spit on them when they came back from Vietnam, but I didn't see any of that. I was hospitalized for a few weeks and the nurses and wives' groups took care of me, and then when I finally got out in town, the people in New Jersey and here in South Carolina treated me decent.

"In Vietnam I didn't worry about politics. When I went to where I was going, it was, okay, here's your job, this is what you do. I didn't worry about what was going on back in the States. I was concerned about getting myself down, doing the things we had to do to keep ourselves alive. If you guys want to scream and holler back in the States about politics, fine and dandy. You have to remember the mindset back in those days was you serve your country, you do what you have to do. Okay, you're sent to Vietnam to fight the communists. That's why we're here. I didn't want to be there, didn't want to get killed, but that's my job. That's why I stayed in the Marine Corps.

"After Vietnam I had a special assignment from 1971 to 1972 as an air marshal at JFK in New York, then went to Japan for thirteen months. I volunteered for drill instructor in '74, went through in nine weeks and became a series gunnery sergeant. Then I was an instructor for about a year. It was unfinished business, you know? I wanted to make up for what I'd done in the past. In 1976 I came back once more, this time as an instructor at the D.I. school. I got picked for first sergeant and went over to the support company under Major Harvey Barnum, who holds the Medal of Honor. He was CO of one of the companies, my company. He and I bumped heads a couple of times but he was straight. He chewed my ass out a couple times. We supported the recruit training regiment, three battalions. The women, who are in the Fourth Recruit Battalion today, were still separate then.

"After that I went to an artillery battery on Okinawa as a first sergeant, then to Camp Lejeune and stayed in the full thirty years. I can't look back and think of anything that I really, really regret. The Marine Corps turned my life around. I tell the kids, 'You know, if it wasn't for the Corps I could be dead or in jail, especially if I had stayed a civilian. All the guys I knew growing up are either dead or incarcerated. The ones who are still out and breathing have no life.' That's what I hear when I do go home and then I say, 'Oh, okay, maybe I did do better for myself.'

"I was guest speaker at a mess night last Tuesday, over at the air station, and one guy asks me, 'Wasn't it boring to spend thirty years in the Marine Corps?' And I said, 'No, I had fun. I enjoyed what I was doing. I mean, where else can you be in charge of a whole bunch of guys or a bunch of stuff, and you're responsible?' As time came for me to retire, the sergeant major called me down and said, 'You're getting pretty short, you know.' I said, 'Look, I'll stay in. You don't have to pay me. Just give me some money for food and I'm happy. I'll do whatever you guys want me to do, you know?' And he says, 'Can't do it. You got your thirty in. You got to go.'

"How is it different from when I was a young sergeant? I tell these guys at PI all the time: 'You're more educated, you know more because when we came everything we learned was basic OJT, on-the-job training. If you had a good sergeant, he mentored you right and you learned the good things and then you were good. But if you got a bad guy, then you picked up all the bad traits. To make rank today you have to go to school, for corporal, or sergeant or staff NCO, all these courses we didn't have back when I came in.' I appreciate that the Marine Corps are finally getting off their butt to see we need education. Because nowadays you can't just take a guy off the street and say, 'Okay, here you are, learn this, learn that.' Sure, you take the rifle, but everybody's not going to be a rifleman. He's got to know something more.

"On the opposite side I don't see that deep down burning in the gut—I wanna be a Marine and serve my country. Lots of guys come in because, well, they like the uniform, that's a sharp-looking uniform, they wanna get the pay, the benefits. If they stay in for four years or six years, there's a GI Bill or whatever."

In comparison with Marines of his generation, Robles does not think the new breed can quite compete with the old. "Physically I think a lot of the old guys were tougher, even though they were not fit like we are now where they could run three miles. But they were harder—they knew physical labor, they knew what it was to work. They were meaner and tougher because a lot of them had to grow up defending themselves in the streets, forty, fifty years ago. In my day, if somebody says something to you, you say, 'Hey, that's it' and you started punching. But it's not like that today and that's why they brought back the hand-to-hand fighting. A lot of these guys have never been in a fight, never been hit.

"The last ten years or so we got to a gentler, kinder Marine Corps, which is what lot of guys like to say. We got a kinder, gentler society so we have to have a kinder, gentler Marine Corps. A lot of people thought we were extremists but, yeah, we have to be, because we're gonna be the guys who are killing the bad guys. We're trying to bring that back in but we don't want to offend the Mothers of America because if they don't send their sons, we're not gonna have a Marine Corps, and if we don't have a Marine Corps, we're not gonna have a country." Robles says that pride in the Corps and in the country is one of the things gaining new emphasis in the Corps. "But we're trying to do it in a way that doesn't offend anybody."

When he compares his experience in training and as a D.I. training recruits, Robles notices a number of differences between today's Corps and the one he remembers. "When I went through boot camp I heard words that you never heard without getting in a fight. But they cleaned that up as time went on, and now you can't call 'em anything. Turds was a favorite because that meant Training Under Rigid Discipline. So you know I can call 'em a turd. But that didn't go over very well either. They're turds, sir, they're trainees under rigid discipline. No, they're not. They're recruits. After a while you could not use profanity toward a recruit. That was in the SOP.

"So you had to come up with creative images. There were other words, you know, like girls, that were not offensive to anyone except a recruit:. 'Sweetheart' was my favorite. My wife hates for me to call her sweetheart to this day, because she knows I used it on recruits: 'Come here, Sweetheart. What's the matter, Sweetheart? Is it your time of the month? All right, girls, we're going outside to do laundry.' You had to watch your language. If you cursed and you were heard, they'd stop your proficiency pay for a month. Well, I probably drew pro pay no more than three weeks all during my first tour. Now you have to go to office hours, and if you get called in too many times, they'll say they don't want you as a D.I. It's rigidly controlled.

"And as for abuse, today that means maybe hollering at a guy or raising your voice. When I talk about abuse back then, I'm talking about hanging a guy out the window with a guy line around his neck that he's holding onto so he won't choke. That's abuse. Did I do that? I'm not going to say.

"And I have to say there were D.I.s who were outright abusive, if not sadistic. They would pick on a guy who rubbed them the wrong way and just keep after him until they drove him crazy or drove him UA, unauthorized absence, which is like going AWOL. And guys would. Parris Island has been cleaned up a whole bunch in the last fifty years, but at one time it was jungle. The little hedge-looking things you see now were all big bushes and you could not see the water, coming across the causeway, because of the oleander. There used to be a steam plant out there, an old brick building, where guys would try to sneak out. Then they'd try to work their way across the causeway, over the bridge. If they tried to swim, they'd get washed out if the tide was strong enough. Then, if they did get across, they had to get through the main gate, and if, for whatever reason, they got past the main gate, then the people in town would turn them in. They knew this guy was a recruit: He's all muddy, he's got utility trousers on, a tee-shirt, and no hair. At that time the police would get fifty bucks a head for turning them in.

"That was way back when, before the main drag was all built up. At that time it was only a two-lane road. A couple police officers, friends of mine, they're out at the Wal-Mart and they see a guy in cammies, walking around the parking lot, looking in cars. So they pull up, call him over, ask for his I.D., ask what he's doing. They didn't realize he was a recruit who had got off the island and he's out there in Wal-Mart trying to find a car with keys in the ignition, so he can jump in and take off. I actually caught a recruit back in my days out where I lived. He was out there taking clothes off a line, trying to get into civilian clothes. I threw a shotgun on his butt, put a rope around his neck, and walked him back to the main gate."

Keeping recruits on the base often is harder than simply having the police look out for wayward trainees. Sometims the D.I.s have to contend with families and other outside influences. "This is a training base, so anybody can come on board if they have a legit reason. A lot of them say they want to visit the museum or they were here fifty years ago and want to go see where they trained and so they're allowed to come aboard the base. The only place I know that's restricted right now is the ammo supply point. Some of these people, usually family members, sometimes help recruits leave the base.

"We try to discourage parents from seeing the kids during training but they will come and say, well, we want to visit Johnny. If it's the weekend when there's no big training schedule, the authorities will allow him to go see the parents, but sometimes the parents will throw him in the back seat of the car and smuggle him out. Then they smuggle him back in, if they don't get caught. Those parents are taking a big chance because everybody in Beaufort knows what's going on, and you got little Johnny out here in camouflage utilities, eating a dinner. It happens the most just before graduation, but I've seen it happen on a weekday or on a weekend during training. Depot regulations say you can't leave a car running with the keys in it, because a recruit might try to jump in and take off. Some of these parents are dipsticks.

"As for the women, I like to tell people I'm the only guy who ever went through the WM boot camp. You don't call women Marines WMs any more. Everyone's a Marine, period. When I was an instructor at D.I. school in 1976, I was given the task of going over to the WM battalion, or the Women's Recruit Training Command, and looking at their training schedule to see how it could be changed or modified to meet requirements. You see, at that time they were in a world of their own, in a separate battalion, and they were taught a certain way and they learned it forever and ever, even though it might be wrong. Drill was taught by a male drill instructor who had gone to D.I. school; but the women never attended D.I. school. They're all OJT. My task was to figure out how we could get women into D.I. school.

"So from Day One, when they got off the bus, I was there until the day they got back on the bus. I was with them. I went along with them every day they trained. I had a notepad, and a tape recorder, and I wrote down how things could be changed or not changed. The only thing I didn't do was take showers with them. Then I gave my report to the director of the D.I. school. But I'm closely associated with them because a lot of things I thought they needed to do they did. And some things I think they shouldn't do, they're doing. For example, they do the hang, no pull-ups. They should maybe do a few. They get rifle training. I didn't believe they should do rifle training. It's a personal thing. I don't think a woman should be in combat. They do need to know how to fire a rifle, though.

"I think women belong as Marines but not as warriors. Sure, you may have to know how to shoot the rifle, but they should be in support. At one time they were forbidden to be in division, they were forbidden to be in an air wing. Now they're in air wing and in the division, a deployable unit. You can end up in combat.

"Regardless how long they're integrated in, a guy will see a woman and do everything he can to protect her and maybe jeopardize the mission or even forget it, just to protect her. Why is she there? It's not the Mothers of America; it's the liberal women who want them there—equality—the right to get killed. Women are nurturers, they're mothers. They're not killers. They protect their babies, they protect their young, but they're not like men, who are the warriors, see? They can be in the miltary, they can be in support. But what can you do? Battle lines are so fluid now there's no front lines like in World War II and Korea. If they're gonna be in support, they should be able to defend themselves, but I don't want to see them running out attacking people.

"A lot of things have evolved, especially where the women are concerned, and we have the Crucible, and recruits do their running in sneakers instead of boots, but we still have our traditions, and tradition has always been a big part of the Corps. Americans love the Marines. They saved us in World War II, they saved us in Korea, they did good in Vietnam, they're doing great in Iraq. So the American public keeps the Marine Corps, even after six guys drown in Ribbon Creek in '56 and two guys die, one a suicide, in San Diego in '76. But the word Marine, capitalized, you know, it's special.

"A kid says, Oh, I'm gonna be a Marine. Wow, that's cool, whereas you say, I'm gonna be a soldier, or, I'm gonna be a sailor, so what? That first guy is cool. He's got high goals for himself. And, when you come down here, they say this is the roughest boot camp there is. Well, I'll be honest with you: It's not really that rough. I've seen other boot camps in the Army that are a little bit more intense in certain areas than we are, but we have that mystique, that thing that says you're a Marine. But there is something about coming here to Parris Island that changes people, especially an eighteen-year-old kid. This is a significant emotional, physical event in a lot of kids' lives. They come face to face with a whole new reality, a whole new way of life. It's either good, or it's bad. You are either going to embrace it or drop out. A thousand a year don't make it, you know.

"But you hear guys who are corporate presidents twenty years later and they look back and they say the Marine Corps gave me direction. They may not really mean that, but it gave them something to look up to, to look back at and say, Yeah, I got it from here.

"I mean, I changed along the way. It's sort of like when I gave up drinking: I started out in Camp Lejeune, and I woke up in Kentucky. That's when I said, 'No more.' I had no idea what had happened in between. That's like five states away. You step on your crank once, that's it. You know I'm not going to step on my crank the same way again. But that was a scary thing on my part—to wake up and not even know where I was, or how I got there. I've been clean and sober since. I was in my thirties then.

"I'll be sixty-four on Monday, April 26, 2004, the same day as my anniversary. That way my wife, Linda Sue, said I wouldn't forget. We got married two years ago."

On the contrary, says Linda Sue, it was he who suggested they get married on his birthday, except she thought he meant they should get married on the Marine Corps birth date, which is November 10. "No, turned out he meant his own birthday."

Robles was married before. Three times. "I have seven girls, five of them adopted, from those marriages. This is my last one. You put thirty years in the Marine Corps, it's a strain on families. But I don't begrudge any of the marriages. I mean, when I look back, it was mostly my fault, all the hell raising and drinking I did."

A regularly practicing Catholic today, Robles shows up every Tuesday from 4:00 A.M. to 5:00 A.M. to help maintain a 24-7 presence in what has become a twelve-year adoration at St. Peter's Church in Beaufort. "I believe," he says, "that I can give one hour a week to Jesus for all the things he has done and given me. Sort of like being on firewatch in the Corps."

Marine Corps Recruit Depot
Training Schedule (Men)

PARRIS ISLAND

PHASE THREE

Recruits are shaping up, so much so that they are now being sent out around the base to function independently at various places in small units. They are preparing to undergo in real uniforms the Company Commander's Inspection and soon after will start Basic Warrior Training (BWT), leading to the Crucible and, after that, graduation as United States Marines.

Training Day 42

Team week consists of Small Unit Leadership among the recruits. The purpose of Team Week is to take them into the next step, from being well-trained recruits to beginning Marines learning small

All that hard work pays off: Graduation Friday on the Parade Deck at Parris Island. *(Photograph by Jean Fujisaki)*

group leadership, which is one of the keys as to how the Marine Corps functions. Drill instructors send small groups of three, four, or five recruits out around the depot to work as a unit in various areas, such as repairing targets at the rifle range, doing the base laundry, working at the recycling center or the base Visitors' Center. Each group has a unit leader who has responsibility for their actions. There is minimal supervision by the drill instructors. Recruits also undergo dental and medical reevaluations during this period. Bruises, sprains, and minor injuries are dealt with from day to day throughout training.

At the end of Team Week, on Training Day 47, comes the Company Commander's Inspection. This takes place in the squad bay. Recruits wear the Service Alpha uniform for fit and serviceability. Recruits were tailored for this uniform, along with desert utilities, on approximately Training Day 20. Recruits pay for all their uniforms. The last week of training they have "final pay bills," in which the recruits repay all their indebtedness with the money they have made throughout the twelve weeks of training. All Marines are paid on the first and fifteenth of each month. Recruit privates receive approximately $1,000 a month.

Sunday afternoons are given over to "remediation," or additional practice at drill, first aid, and other classes.

Training Day 48

This week is devoted to Field Training or Basic Warrior Training. Recruits are bused to various sites to learn rappelling, take a day movement course, practice unknown distance firing, and take a combat endurance course, consisting of various obstacles on a three-mile track, performed in boots and utilities (boots and utes). In unknown distance firing, recruits will fire at targets spaced at odd distances. In MOUT, a movement course in urban training, they learn how to maneuver in and around buildings.

Training Day 54

Recruits undergo their final fitness test, the PFT, consisting of three pull-ups, fifty crunches, and a three-mile run in shorts, tee-shirts, and

Go-Fasters. This must be completed in twenty-seven minutes or less. The maximum number of pull-ups is twenty, crunches can go to a hundred, and eighteen minutes is the maximum for the three-mile run. Women do a seventeen-second flexed-arm hang and fifty crunches, and they have twenty-nine minutes to run three miles.

Training Day 55

On Tuesday at 0200 (2:00 A.M.), on the tenth week of training, the Crucible begins with a six-mile hike to Page Field. Recruit platoons are broken into three groups with the focus on small unit leadership. There are six events, each with eight stations replicating some aspect of combat activity. They are structured in a way that requires group problem solving and interactivity to accomplish a goal. Recruits will hike back Thursday morning after two and half days. They have had four hours' sleep each night under the stars in sleeping bags. Each gets MREs, or Meals Ready to Eat. They are given four in the wintertime and three in the summertime, and they must ration their intake for the three days. They can eat any time they get a break. The Thursday morning hike covers nine miles back to Leatherneck Square. At the conclusion of the Crucible comes the Warriors' Breakfast, an all-you-can eat buffet shared with the drill instructors. Recruits are so excited by the food "they barely look up from their plates," Staff Sgt. Wylie said.

After breakfast recruits return to squad bays to clean their rifles and prepare for the last few weeks of training, turning the rifles in and getting ready for final drill and written tests. Rifles are turned in after the Battalion Commander's Inspection on the Tuesday before graduation, which takes place Friday. Inspection occurs on the Parade Deck.

Altogether, six hundred drill instructors train sixteen companies of six platoons each simultaneously aboard the base, roughly 6300 recruits, including the women. Plans were to reach 7700 and 8000 in the latter part of 2005. Platoons may be as small as 37 in the wintertime and in the summer, following high school graduations, platoon numbers swell as high as 87. Parris Island trains 16,000 Marines annually, 2400 of whom are women.

Week Eleven

Training Day 60

This is testing week, when recruits take a 171-question written examination, covering all academic instruction, Corps history, customs, and courtesies. The rest of the week includes a final drill evaluation on Wednesday; a practical application exam, focusing on hands-on demonstration of first aid activities; more classes; graduation practice; and drilling out on the Parade Deck.

Week 12 is an administrative week, for the most part. The Battalion Commander's Inspection comes on Tuesday. Rifles are turned in that afternoon. In fact, recruits will return everything they received—their bucket issue, their ALICE pack—while their personal effects, civilian clothes and the like, will be returned to them. Recruits will also receive orders to their next duty station, which will be Marine Combat Training or School of Infantry at Camp Lejeune. Recruits receive the Eagle, Globe and Anchor insignia in a ceremony on the Parade Deck on Thursday. Families are invited. Family day follows, in which recruits can spend from 1300 (1:00 P.M.) to 1900 (7:00 P.M.) with their families. Generally, it is the first they have seen them since training began. Graduation occurs Friday, also on the Parade Deck. Families and friends observe the ceremony from bleachers.

Prior to graduation, a senior drill instructor, such as Staff Sgt. Wylie, will address the platoon informally in the squad bay. He—or she—will note that recruit training has been extremely stressful and note that the fact they have chosen to enlist during wartime is commendable. He or she will say that their completion of a challenging training process at such a time speaks highly of them all. The title of Marine makes their last thirteen weeks only that much more special. "It gets emotional," Staff Sgt. Wylie said. "They put their heart and soul into every moment they have been here. They have been changed from what they used to be into a smartly disciplined Marine, and they now see the difference. They see that what they have done for the past thirteen weeks will carry them forward the rest of their lives. It is very emotional, yes, sir.

"It's amazing the changes they make from day to day and even moment to moment. From having worked in Receiving and taken recruits off the Yellow Footprints all the way to Training Day 70 on graduation day, it's hard to explain how various events in the training process mold them and take away the poor habits they had when they came in. My platoon now is on Training Day 18, getting ready to go to Initial Drill tomorrow. Only three weeks into recruit training, they've already started to function as a team. You see the teamwork between that recruit who is a little weaker in a certain area, you see two other recruits going to help him to see that he successfully completes a task. And it's only been three weeks. They've got a different look about them. They're physically changing. Their bodies are becoming slimmer and more muscular just in a matter of weeks. It now becomes a matter of refining and accumulating skills as they continue through training."

CHAPTER SIX

ROBERT MASTRION

Mustang
1958–1992

For combat leaders, the images of the
Marines who were killed or wounded never leave you.
The minute your mind is not focused on the present, these images
rush in and grab your focus. The images never go away. Your mind can
never wander anywhere else; these images always fill any void
in your focus. The older you get and the less there is to
hold your concentration, the more these images
are right in front of you.

Twice court-martialed, twice awarded the Silver
Star. In 1968, Colonel Robert Mastrion was a captain
commanding Company G, Second Battalion, Fourth
Marine Regiment, operating just below the DMZ
(demilitarized zone) in Vietnam as part of the Third
Marine Division. In this photo, taken in 1969, Captain Mastrion has just been promoted to major. *(Photograph courtesy of Robert Mastrion)*

Robert Mastrion enlisted in the Marines 1958 and made full colonel in 1983. While he never became a drill instructor, he did undergo recruit training at Parris Island and eventually became a mustang, meaning he rose through the ranks and attended OCS (Officer Candidates School). As an enlisted man, he was court-martialed twice, for fighting, and as an officer he was awarded two Silver Stars while leading his men in combat during one and a half tours in Vietnam. He also received two Purple Hearts. He does not remember what the first Silver Star was for. The citation for his second appears at the end of this chapter. He spoke to me over the phone from his home in Mount Pleasant, South Carolina.

When I first spoke with Col. Mastrion in the spring of 2005, he immediately began discussing the best way to form a functioning Marine Corps and explained how the Corps encountered problems with personnel at the end of the Vietnam era. "There's a reason why we can consistently turn out good fighters in the Marine Corps and that's because we eliminate the problem children right off the bat," he said. "But you gotta understand human nature. For example, take a piece of paper and draw a line across the page midway down. Then draw a vertical line right down the middle. Mark 50 percent at the point where the two lines intersect. Then mark 60, 70, and so on in 10-percent increments up to the top. Mark 40, 30, 20, and 10 going down from the center point.

"Now, in any organization or group or kindergarten or college classroom or business, you've always got about 3 percent that are incorrigible son-of-a-bitches, that's from 0 to 3 percent. And there is nothing, absolutely nothing, that you can do with these people because they're just selfish, absolutely selfish. Think back to your days in high school. You always had people getting in trouble but it was usually a real small group of guys—or usually just one—that would always be doing things to get in trouble. You just have to leave these people alone. I ran a business here in Charleston after I retired, and I'm now retired from that, but the one thing I always looked for was the 3 percenters.

"And from 3 to 10 you've got what I call the 7 percenters. They are essentially the same as the 3 percenters, except they are not as active. They're more like followers, but, given half a chance, they'll become 3 percenters. And from 10 to 20 percent, you've got people who tend toward being 7 or even 3 percenters, but they're mostly passive. And if the 3 percent get control enough to influence the 7 percent, then the next 10 percent will move that way also. That's always with you, so you've got to watch that.

"At the other end of the group, at the top, you've got about 20 percent who will do no wrong. They'll do good no matter what.

"As for the 60 percent in the middle, they'll go whichever way looks like will benefit them the most. So you're in a constant battle for that 60 percent.

"Now, in the Marine Corps, the 3 percenters are usually weeded out by the recruiters, who also reject most of the 7 percenters. The recruiters watch this like hawks, because, if a recruit doesn't make it through boot camp, they get nailed, they get a black mark on their records. We say we are looking for the few good men, but what we're also doing is picking out the bad guys and eliminating them. So right off the bat we have an advantage that a draft-driven army never has, because we chop off those 3 and 7 percenters.

"We still have 8 or 9 percent attrition in boot camp. The turds are the guys from 10 to 20 percent and we weed them out in recruit training. Demands for a spirit of cooperation and unselfishness weed out the 10 to 20 percenters, for the most part. The goal is to eliminate that whole cadre, because that is what undermines an organization.

"The recruiters aim to wipe out the first 10 percent before they even get on the plane. Now, one may slip through every now and then, but, boy, they don't last long. It's not like when I first went in. Hell, you walked in and they were happy to get you. But then again boot camp was different, plus the individuals going in were different, because our society wasn't as self-centered as it is today.

"If the problem kids get through and get out in the Fleet Marine Force, you've got problems. Speak to anyone who was around in the 1970s after the draft ended in '75. I wasn't in recruiting and I wasn't at the Recruit Depot but I was in the FMF at that time when the draft went away and it got difficult to recruit because we still had large forces.

"I was down in the Second Marine Division and the barriers against the 3 and 7 percenters had been done away with because of all the people who were being brought in. They were just grabbing anybody they could, and the Fleet Marine Force unit I was in was paralyzed because we had so many of these bad guys running around. And you don't need many to paralyze a unit. If you hit 13 percent, the unit starts to come unraveled. If you got 15 percent of these guys, you're paralyzed. I don't care if every officer and every staff noncommissioned officer was a John Wayne," said Mastrion. "You still won't have a functioning unit.

"The Marines didn't have that many problems in Vietnam, not half as many as the Army had. But the barriers were still there. The draft was driving them toward us but we were still selecting who we wanted. The barriers went down when they did away with the draft. And, once those barriers were gone and we started to fill up with these guys, that's when we started to have problems. This was after the Vietnam war. The pressures to fill up, to meet quotas, was so great and the impetus wasn't driving recruits to any of the services so the pressure to fill up was so great that the 3 and 7 percent barriers went down. We were out I think 200,000. So the barriers went down and these guys started to come in. And again you don't need many because, remember, the 3 percenters will immediately ally with the 7 percenters."

Mastrion does not have kind things to say about the quality of recruits in the early 1970s. "I got to the Second Division in 1972 and it was bad then, and it wasn't until the end of '74 or the beginning of '75 when General Wilson took over as Commandant that things began to change. The whole Marine Corps had been having problems with these turds, and it was just difficult to get rid of them because of the need for people. We were paralyzed. You couldn't train, you couldn't do hardly anything because you were taking care of these people at the low end of the bell curve. We had had discipline problem after discipline problem after discipline problem. I remember that one week I was with the Third Battalion, Sixth Marines, in one week, the UA [unauthorized absence] rate broke [went under] 100, and we were happy. I mean we celebrated. It was a very bad time and the reason why the UA rate was so high was that the good guys were saying the hell with this and going over the hill.

"We lost a good cadre of staff noncommissioned officers, which essentially is the backbone of the Marine Corps, and it just was terrible. The daily rate was over 100 (out of about 900 on board), and we were happy that it went below 100. It was a very bad time. Many of the UA Marines were good men who were tired of the criminals who were making their lives miserable. Our joy in having "less than 100 UAs" was a bit tongue in cheek. Breaking 100 really meant nothing, as things did not get better for some time to come. Units were tied to garrison—meaning you couldn't go to the field—because they had so many disciplinary problems. It appeared that the powers that be in Washington, D.C., were obsessed with keeping the Corps numbers up, and it was very difficult to discharge troublemakers.

Captain Mastrion's Company G had captured a Vietnamese village just before this photo was taken in 1968. *(Photograph courtesy of Robert Mastrion)*

"Then in 1975 Lt. Gen. Lou Wilson was selected to be the new Commandant. Before he was installed, while still commanding Fleet Marine Force Pacific in Hawaii, he started sending messages to the Commandant, with information copies to every major command in the Corps, outlining how he was addressing the issue of getting rid of problem Marines and what the Commandant should do to speed this process. This was taken as a signal by everyone up and down all chains of command, to include Headquarters in Washington, that the time had come to purge the Corps of bad apples.

"I got home from the Mediterranean June of 1974 and we started to get rid of these incorrigibles through administrative discharges left and right. The number of administrative discharges handed down was staggering. Once that started, you could look out the window and see troops on the road going out to train. So you can see what happens if you don't keep out the 3 and 7 percenters and don't weed out a good portion of the next 10 to 20. I saw this as validation of the theory."

But Mastrion thinks the problem runs much deeper than the Marine Corps. "Remember *Seinfeld*? That show was brilliant because it showed normal people who were so self-centered they were obsessessed. Remember the last show, when all those people came up and told them how their self-obsession caused all that pain for all the rest of the people? Remember that? That was a good measure of how society had changed. With *Seinfeld*, nobody really noticed how selfish those people were until that last episode. I know a lot of guys didn't like that last episode because it hit them right in the face."

Seinfeld, he believes, is symptomatic of society in general these days. "Remember how, when you were a kid, the worst thing anybody could say to you was, 'You are selfish'? Remember that? You'd go around doing good deeds for a week until somebody said you weren't selfish. To my way of thinking, our whole society grew self-indulgent. The World War II era was a different world. People were different. I mean, Jesus, my dad was making thirty-eight bucks a week and you didn't have all this showering of pleasure on people.

"Kids today, recruits even, are more self-centered. You can see, like, even in the grandkids. I'm probably the cause of it because I spoil them, but, in any event, our whole society has changed to where this has to be coped with. In boot camp today, they have to concentrate harder on getting the selfishness out of these kids.

"By emphasizing tradition, unit cohesiveness, harder discipline, being harder on them physically and psychologically, we're able to mold them quicker. The focus is on confronting lax and self-indulgent behavior in personnel.

"Back when, and up through the '70s, the material was there. All you had to do was mold it. But, as our society got more self-centered, they even had to screw with the material to harden it up, and they did that rather well, I think. The fact is, we put out a pretty good product consistently even though the base product has changed.

"The mistake that was made in our society was we started to concentrate not on changing the 3 and 7 percenters, but accommodating them. That's why all these kids—we've got schools falling apart. But the genesis of that falling apart is actually the 3 percenters and the 7 percenters.

"I have followed this and thought about it at length, even though I never served on the drill field at Parris Island. I was only a recruit there. I was an enlisted man for six years, which was when I got court-martialed twice. I joined the summer of '58, after Ribbon Creek, a year after I got out of high school. I'm sixty-five. My birthday is July 30.

"I grew up by the Navy Yard section by the Williamsburg Bridge in Brooklyn. I came into Yemassee by train. There was a Receiving Barracks at Yemassee up on a hill. I remember what they did to us was brilliant. We got in there, and, as you got closer, the conductor started to collect all the recruits into one car. You could see the process start once you left Washington. We had a bunch of guys from Buffalo, New York City, Philadelphia, Baltimore, a lot of tough kids.

"They got us down to Yemassee all in one car at eleven o'clock at night, 2300. A D.I. got on the train and started screaming and hollering, 'Get off! Get off! Get off!' and all that. One kid went into the bathroom and locked the door. Without even losing a step the conductor took a screwdriver out of his pocket and took the hinge off. I saw the kid run into the bathroom and I saw him dragged out later, so I know they got him out.

"They lined us up right on the side of the train, shoulder to shoulder, and made us put our noses on the train, just lean forward, and then, as the train started, they made us take one step backward. They made us stand there as the train started to pull out, and they just left us there. I guess the conductors and the engineers were in on this because they kept blowing the whistle and we just stood there for, I don't know how long, until the whistle got very faint, 'Whooooo.' You can imagine the psychological impact of that. Then they ran us up to the recruit barracks where we waited about three hours until the school buses came to drive us over to PI. They made us put our heads between our legs so we couldn't see how to escape. I mean, how stupid is that? It's dark out. But they did it and got us into Receiving and ran us to bed.

"Boot camp was tough but there was no brutality. I knew about Ribbon Creek, I had read about it in the newspapers, and I think the trials were going on. They always managed to discipline you even though everybody was being watched. I was a wise ass and I got my comeuppance. I got whacked in the face a couple of times to correct my position and to correct my attitude but we were never beaten. They would put a hand on your shirt collar then accidentally whack you on the chin as they took it off. But it wasn't a right cross or anything. They just made their point. But, remember, kids were a helluva lot tougher then.

"I'll tell you one thing that bothered the hell out of me and that was holding an M-1 rifle by the stacking swivel. The M-1 has that little stacking swivel that hooks them together when you stack them, and the drill instructors would make you hold your rifle out by that swivel at arm's length in front of you. You were holding it out there with two hands, arms extended all the way. The M-1 weighed 9.5 pounds. That helps you learn. You learn to pay attention. They had a lieutenant, a series officer, who oversaw four platoons.

"I completed boot camp in the fall of '58 and went to Camp Lejeune as a mortarman. After that we went on cruises, to the Caribbean, to the Med, and they sent you to Okinawa. I enjoyed it. As an enlisted man, I got demoted to private twice. But I worked hard at it. It was mostly for fighting. I never went UA. That was unheard of then.

"I applied for OCS toward the end of my enlisted tour. I had a GCT of 136, which is like your IQ, I guess, and when Kennedy got in they were expanding the services. I got called in and they said you got a 136 IQ, do you want to go to OCS? I said no. But they kept talking to me and finally I said okay.

"Officer Candidate School is physically tougher than boot camp. At Quantico you got hills. It's called the hill trail and they ran us across that, God, you just kept running and running up and down these hills. It went on for twelve weeks. Kennedy was in already and that's when I got a guy in trouble because I was a wise ass. They got a new Secretary of the Navy and we were getting ready for a big inspection and somebody came running in all panicky, asking, 'Who is it? Who is it?' And I said Peter Lawford, that actor who was married to a Kennedy. Well, the colonel came through and asked that very guy who the new Secretary of the Navy was, and the guy said Peter Lawford. The platoon sergeant was standing next to him and, Oh, God, his face just melted. I went in later and admitted I had told the guy that but said I thought he knew I was joking. They said get out of here.

"I got commissioned December of '61. I went to Vietnam the first time in '63, on a rotation out of Okinawa. Then I went again, from May '67 to June '68, a little over twelve months. I was the company commander. I was in Second Battalion, Fourth Marines and I had Golf Company for a while. It's so foggy I can't really remember half the stuff. It was a good unit. We were located up by the DMZ facing North Vietnamese Army units. We were out all along there—Quang Tri, the Rockpile, Camp Evans, Route 9, Con Thien, places like that. We were pretty busy.

"My wounds were from a combination of things. You usually don't get just one thing, you get a bunch, shrapnel fragments, bullets, things like that. It's no big deal. You know you got no choice. Getting wounded's easy. Getting better's hard. They'd patch us up and send us back out. In one instance, they needed help, so a bunch of people got out of the hospital and went back in, we just got up and left. You're okay. Life is arduous. There were no heroes. You just did your job."

In 1969, Robert Mastrion was promoted to major, and, in 1978, he became a lieutenant colonel. He retired a colonel on July 1, 1992.

As a captain, Mastrion received two Purple Hearts and two Silver Stars during his tour in Vietnam.

CITATION (SILVER STAR)

For conspicuous gallantry and intrepidity in action while serving as Commanding Officer of Company G, Second Battalion, Fourth Marines, Ninth Marine Amphibious Brigade in connection with operations against the enemy in the Republic of Vietnam. On 30 March 1968, Company G was assigned the mission of attacking the village of Nhi Ha in Quang Tri Province. When adjacent units became pinned down by intense enemy fire, Captain Mastrion was forced to employ his company in an area exposed to a heavy volume of artillery and automatic weapons fire. Boldly exposing himself to the hostile fire, he quickly organized a defensive perimeter, despite the accurate enemy fire which knocked the radio handset from his hand on two occasions. Completely disregarding his own safety, he fearlessly continued to move across the fire-swept terrain, shouting words of encouragement to his men and ensuring that they were occupying positions of relative safety. Although wounded twice, he steadfastly refused medical treatment and continued his determined efforts for approximately twelve hours, until his unit was ordered to withdraw from the hazardous area. His superb leadership and aggressive fighting spirit inspired all who served with him and were instrumental in the accomplishment of his unit's mission. By his courage, bold initiative and steadfast devotion to duty in the face of extreme personal danger, Captain Mastrion upheld the highest traditions of the Marine Corps and the United States Naval Service.

> For the President,
> H. W. Buse Jr.,
> Commanding General,
> Fleet Marine Force, Pacific

CHAPTER SEVEN

EDDIE ADAMS

Staff Sergeant
1951–1954

I once photographed Chesty Puller.
Another one of my first assignments, after I
got into the Photo Section of the Second Marine Air Wing,
was to take a picture of Ted Williams climbing into his plane.
This was for the base newspaper, *The Windsock*.

Eddie Adams with his son, August, in their New York City home, in a photo taken by the author several weeks before Eddie died of ALS at the age of seventy-one on September 18, 2004.

Eddie Adams, to whom this book is dedicated, died of Lou Gerhig's disease, amyotrophic lateral sclerosis, on September 18, 2004. He was seventy-one years old. Over a forty-five-year career he covered thirteen wars and won hundreds of photojournalism awards, shooting for Parade *magazine,* Time *magazine, and the Associated Press, among others. It was for the AP that he took one of the best-known photographs of the twentieth century, when he caught the exact moment that Brig. Gen. Nguyen Ngoc Loan, the national police chief of South Vietnam, fired a bullet into the head of a Vietcong prisoner on a Saigon street, on February 1, 1968. The picture won a Pulitzer Prize and was considered instrumental in ending the war, but Eddie said it failed "to depict the general as a great soldier, a great, warm human being, loved by all, and a good family man." He said two people died in the photograph. The man who pulled the trigger was the other one. Eddie began his professional career as a Marine and, like any good Marine, he had a bawdy sense of humor. We talked about his experiences in the Corps not long before he died. He was a good friend.*

"I joined the Marine Corps in July 1951, after graduating from high school in New Kensington, Pennsylvania, near Pittsburgh. I had already been working nights for a year, taking pictures of accidents, sports, ballgames, and PTA meetings for a local paper. I was eighteen years old when I went in.

"We were assigned to Platoon 321, Second Battalion. We had two drill instructors. One of them was a staff sergeant, a tall, thin southern boy with a thin mustache. He was fairly pleasant. The second D.I. was about five feet six, with a rugged, nasty-looking face. He had already served eleven years in the Corps. He was the bad guy, but he also was the one who made real Marines out of us.

"He used to stand in front of us talking and, at the same time, he would peel warts off his hands with a jackknife. His hands were bloody. He would do it like peeling an orange.

"I remember one night about 0300 he heard somebody talking. He got us all up out of our racks, called us to attention, then had us pick up our lockers, heavy wooden boxes that held everything we owned. First he made us do left and right shoulder arms with them. Then we had to get in formation and run up and down two levels of stairs. While we did this he would shout orders for left and right shoulder arms. Several of us fell on the steps. He screamed that if somebody fell, the others should run over them, which they did.

"Another time a little guy was caught smoking. The corporal called everybody to formation in the squad bay and then had the little guy sit on a chair on top of table. Then he had somebody put a full pack of lit cigarettes into his mouth. Then he put a bucket over the guy's head. Then he put a poncho over that, and a blanket over the poncho. We all stood at attention as the D.I. screamed that he wanted to see more smoke coming out from under the blanket. The recruit later was taken to the hospital. The D.I. was not punished.

"Women were not very welcome in the Corps at that time. You didn't see them that often. The word was that most of them were ugly and they only joined the Corps to get laid. I don't know if that's true but I never forgot being out on the drill field one day and hearing a WM D.I. scolding her platoon, whch was right behind ours.

"She was yelling at her recruits, and she said, 'There are ten miles of cock on this island and you people are not going to get one inch of it if you don't you get your act together!' I never forgot that.

"Sand fleas were our biggest enemy on Parris Island. They used to bury themselves in us, especially when we were in the prone position at the rifle range and not allowed to move.

"That's when I wanted to die. I didn't think I would be able to complete the training. I didn't think I would make it.

"After graduating from boot camp, I was assigned as an air controller with the Second Marine Air Wing at Cherry Point, North Carolina. I asked to speak with the commanding officer and requested reassignment as a combat photographer. I spent a year in Korea, starting three months before the armistice was signed. We used Speed Graphics in the service at that time, then I went to the 35-mm Nikon. I once got a writer court-martialed for taking him to a whorehouse in Seoul.

"I got out of the Marines in July 1954 and spent three more years in the Reserves. The rest is history.

"I went to work for the *Daily Dispatch* in New Kensington, then the *Inquirer and News* in Battle Creek, Michigan, then the *Philadelphia Bulletin* from 1958 to 1962, when I joined the Associated Press. I was there ten years, worked for *Time* magazine from '72 to '76, went back to AP as a special correspondent, then to *Parade* from 1981 to 2004. That's how my life happened."

PART TWO

RIBBON CREEK

Sea Change in the Corps

THE CORPS was sailing along, riding high on its performance in the Pacific during World War II and in the Korean
conflict, until Staff Sgt. Matthew C. McKeon led Platoon
71 on a night march into Ribbon Creek on Parris Island.
Six men drowned, and the court-martial that followed
was sensational. It brought the training techniques of the
vaunted Marine Corps to light and for the first time pub-
licly exposed a D.I. to intense national scrutiny. As Morton Janklow notes subsequently in Chapter 9, it was
perhaps the second-largest event of its kind, after the
trial of Billy Mitchell, who tried in vain to convince the

brass that air power was going to be important in the wars of the twentieth century.

As mentioned in the Introduction, and as is clear from the stories of the Old Breed, training methods had grown seriously out of hand. Now, the Ribbon Creek incident had put the Marine Corps itself on trial. Abuse of recruits, though specifically prohibited, was commonplace. Examples of maltreatment included "thumping," punching recruits in the stomach, burning recruits with cigarettes, forcing them to eat the butts, stacking them in trash bins, and making them run a "belt line," which was a gauntlet of belt-swinging fellow recruits. Acts that were later considered hazing included duckwalking (being compelled to walk while crouching with knees fully bent), hiking with packs full of sand, dryshaving while running in place, and using locker boxes as barbells.

Drill instructors were NCOs, sergeants primarily. Corporals also served as D.I.s. Occasionally, privates who performed well in recruit training were granted the privilege. There was minimal officer supervision. After Ribbon Creek, all this was supposed to change immediately, despite strong resistance in the noncommissioned hierarchy. A separate Recruit Training Command was established at Parris Island, to be commanded by a brigadier general selected by the Commandant of the Marine Corps and reporting directly to him.

As previous chapters of this book indicate, NCOs did not altogether give up thumping, and other harsh training techniques, even though they were prohibited. Some D.I.s who exceeded training limits ended up in the brig. Nevertheless, the next several years saw a sea change in D.I. culture while the training system was being overhauled. For example, three instructors were now assigned to each platoon, instead of two, each D.I. was to receive extra pay of thirty dollars a month in hopes of relieving stress as well as financial strain, and all training was closely supervised by officers looking for ways to improve procedures. Drill instructors were urged to emphasize training by example, persuasion, psychology, and leadership, rather than browbeating. They were exclusively authorized to wear the campaign hat (along with Distinguished Marksmen at the rifle range).

In addition, a special unit was set up to deal with the disobedient, as well as recruits who had specific problems. A conditioning platoon was created to provide special diet and exercise to deal with the overweight, a proficiency platoon was set up for slow learners, and a motivation platoon was put in place to deal with recalcitrants.

It can be said that a great deal of maltreatment grew out of the fact that the drill instructors were expected to make Marines out of some who were basically incorrigible— misfits or bad apples. The extreme training techniques evolved, in part, from the frustration of trying to train difficult recruits. The Motivation Platoon was intended to ease this problem, but even that did not work in the long run. Ultimately, after the Vietnam War, the Corps finally adopted a policy of returning misfits to civilian life quickly so the D.I.s would not have to waste time with fundamentally untrainable individuals.

CHAPTER EIGHT

MATTHEW C. McKEON

Staff Sergeant
1947–1959

Some swear by you; some swear at you.

Staff Sgt. Matthew C. McKeon led his platoon on a
night march that ended in death for six recruits.
Gen. Randolph Pate, at the time Marine Commandant, said McKeon "positively did not have the
authority" to take the men on the march. McKeon,
of Worcester, Massachusetts, was thirty-one when
this photo was taken. (*Photograph © Bettmann/
CORBIS*)

If the Marine Corps has a tragic figure, it is Staff Sgt. Matthew C. McKeon, who died in relative obscurity at the age of seventy-nine on Veterans Day, November 11, 2003, in West Boylston, Massachusetts. He had lived there forty-one years.

McKeon, of course, was the drill instructor who led the seventy-five recruits of Platoon 71 on a dark, moonless march the night of April 8, 1956, into a muddy backwater called Ribbon Creek. Six recruits drowned when the platoon panicked as it encountered an unexpectedly strong tidal current in an area of deep pockets that was unknown to McKeon.

"They were all good kids," McKeon said forty-two years afterward, in an interview with the Long Island newspaper *Newsday*. "I wasn't really angry at them. And hurting them was the furthest thing from my mind. I was trying to discipline them."

During the court-martial that followed, McKeon was found not guilty of the more serious charges of manslaughter and oppression resulting in death. Allegations of drunkenness proved unfounded, although McKeon acknowledged having a few slugs of vodka with another drill instructor earlier that day. He was found guilty of involuntary manslaughter by simple negligence and of drinking in an enlisted men's barracks. He was reduced to private, sentenced to nine months of hard labor, lost thirty dollars per month in pay, and was given a bad-conduct discharge.

But the penalty was reduced by the Secretary of the Navy, Charles S. Thomas, to three months in prison and reduction in rank to private. The bad-conduct discharge was pardoned, and Thomas said, "For him I believe that the real punishment will be always the memory of Ribbon Creek . . . remorse will never leave him."

McKeon was transferred to the Marine Air Station at Cherry Point, North Carolina. Not long after, he was promoted to corporal and even named Marine of the Month. After twelve years' service, he left the

Corps in 1959, with a medical discharge stemming from two ruptured disks in his back. He went home to West Boylston and drove a milk truck for a while before landing a state civil service job as an inspector. He held that job thirty years before retiring in 1990.

In the *Newsday* article, published in August 1998, McKeon was quoted as saying Secretary Thomas "was right," and he added, "Living with the memory has been a far worse punishment than anything they could have done to me. At times I wish I went down with them. I don't think any judgment or any justice could be severe enough to pay for those kids. I don't think I'll ever get what's coming to me. All I can say is that I'm truthfully sorry."

The following year, August of '99, McKeon told a reporter for the *Worcester Telegram & Gazette* that not a day went by that he didn't think of "the incident."

Many people in the Worcester area knew about the drownings when McKeon and his wife and five children moved to West Boylston in 1961. "A lot of people helped," he said of the reception he received at the time. "If it wasn't for fellow Marines, friends, people in the neighborhood . . . I don't know where the hell I'd be. I'm thankful for these people. I wouldn't have made it without them." But then he added, "Some swear by you; some swear at you."

The story of Ribbon Creek has been explored and recounted in detail in *The U.S. Marine Corps in Crisis: Ribbon Creek and Recruit Training*, by Keith Fleming, published in 1990, and in *Court-Martial at Parris Island: The Ribbon Creek Incident*, by John C. Stevens III, published in 1999. Both writers state unequivocally that the atmosphere at Parris Island, methods of training recruits, and the free hand given to the noncommissioned officers who served as drill instructors all combined to make the drownings an accident waiting to happen.

The newly retired Lt. General Lewis B. (Chesty) Puller, a Marine Corps icon, testified during the court-martial, in response to a question from McKeon's lawyer, Emile Zola Berman, that it was "good" military practice to lead men into water for the purpose of instilling discipline and morale. He did call McKeon's night march a "deplorable accident." But when the prosecutor asked the revered general about taking action "if I were the one who did that, if you

A troubled McKeon, hatless and slumped, awaits the outcome of his court-martial in the drowning of six recruits in Ribbon Creek at Parris Island in April 1956. *(Photograph by Robert W. Kelly/Time & Life Pictures/Getty Images)*

were my commanding officer," Chesty Puller replied, "I think from what I read in the papers yesterday of the testimony of General Randolph McPate before this court, that he agrees and regrets that this man was ever ordered tried by general court-martial." Puller held five Navy Crosses, one decoration shy of the Medal of Honor, and he was the most highly respected Marine in the United States. His statement was a bombshell. The defense rested, and McKeon was found not guilty of the most serious charges.

But he could not get out of his own skin, and thirty-three years went by before he finally broke his silence. In November of 1989, speaking to Wil Haygood of the *Boston Globe*, McKeon said, "I think about the boys every day. I do. Oh, God. You run these names over and over in your mind. I wonder what kinds of jobs they would have, if they'd be married, if they'd have kids." Their names were Norman Wood, Charles Reilly, Donald O'Shea, Jerry Thomas, Thomas Hardeman, and Leroy Thompson.

"The court was right. They convicted me. I got no problem with that. No, sir." He acknowledged taking "a drink down there," and said "everybody," meaning the drill instructors, drank.

Platoon 71 was McKeon's first group of recruits. After service in the Navy in World War II, he enlisted in the Marine Corps in 1948 and served in Korea. He had graduated from drill instructor school in January of 1956. He was thirty-one years old at the time. The platoon was shaky, and McKeon's aim was to instill discipline by means of the march. One recruit later said McKeon had told his recruits, "Now, the ones who can't swim will drown. The ones who can swim, well, the alligators will get you. So it won't make a damn bit of difference."

McKeon told the *Globe* reporter in 1989 that he was not an especially religious man, but, he said. "I believe in God. There'll come a Judgment Day. I'll have to answer."

He added, "Sometimes, the living have it harder than the dead."

CHAPTER NINE

MORTON JANKLOW

Assisting attorney,
court-martial of Matthew C. McKeon
June–August 1956

If I were in combat, what I would want
next to me in the foxhole is any Marine drill sergeant.
Not an Army commando, not a Ranger, a drill sergeant. Those
guys, first of all, could go for thirty-six hours without a wink of sleep.
They were amazing guys. They worked like dogs for nothing. I mean,
they got paid nothing and in that heat. They were up at four
in the morning. They didn't go to bed till midnight. And
he [McKeon] was just not cut from that cloth.

Morton Janklow, a young army veteran, was
tapped by Emile Zola Berman to help provide
a defense for Matthew C. McKeon. In his
hands is a clipping from the Sunday, July 17,
1956, edition of the *New York World Telelgram*
showing himself, Berman, and a despondent
McKeon at the trial. (*Photograph by Gina Levay*)

Morton Janklow is a prominent New York City attorney *and literary agent who, because of his prior experience in the Army with courts-martial, was asked by Emile Zola Berman to assist in the defense of Staff Sergeant Matthew C. McKeon. Janklow had been a law school friend and colleague to Howard Lester, who worked in Berman's office in New York City. Berman himself died in 1981.*

"Sergeant McKeon lost these six guys and it became the biggest court-martial in American military history since the Billy Mitchell case. The whole national press corps was down there. You can't imagine what it was like. It was like the Michael Jackson trial in the military.

"The Marine Corps had immediately decided that this guy was a threat to the Corps and to the public perception of the Corps and so they were going to make him an example. A relative, a judge in New York City, was very, very concerned that McKeon was going to go to jail for a long time. The judge asked Emile Zola Berman to represent McKeon.

"Everybody [in New York] knew Zuke Berman. He had a huge reputation. He was a negligence lawyer. He was a very human guy. He looked like a caricature by the French artist Honoré Daumier: He was skinny, he had a long, beaked nose, beady eyes, and he liked to drink.

"Berman and I and a guy called Howard Lester spent four or five months of our lives, uncompensated, representing him because we thought there were national issues at stake: How you train men for combat. We thought it was a very important court-martial. Berman and I both agreed that the Marine Corps was railroading this guy. They were scapegoating a guy who did nothing out of the ordinary, he just had bad luck. And they were really laying it on him. If you looked at the Marine Corps PR at the time, it was, 'Oh, this is unheard of in the Marine Corps training' and so on. But the fact was, the Marines had suffered fewer combat casualties per hour than the Army in Korea because they were better trained. That was true in the Second World War too.

"The court-martial is not designed for the administration of justice, by the way. The Uniform Code of Military Justice is not about justice; it's about the administration of discipline. A defendant facing a general court-martial could pick anybody he wanted to represent him.

"I was a kid, twenty-six years old. I didn't have any money. We're talking about 1956. I even had to drive down to South Carolina because the airfare was a big deal to me. And we get down there and Berman and I are totally isolated. The trial hasn't started, we haven't picked a court, and we can't get to first base. I'm walking around just observing the troops drill training and I'm watching and I remember the first thing I saw occurred when I wasn't on the base an hour. There was a guy drilling some kids behind a barracks and I went over there. I wanted to see what the course of normal training was like. I was the leg-man.

"There was a kid, some bohunk kid from Mississippi or someplace who didn't know his left foot from his right, kept marching badly. The drill sergeant walked over, very cordial and sweet, and said to him, 'Are you having trouble, keeping the rhythm of the march?' And the kid said, 'Yes, Sergeant.' He said, 'Well you seem to be having trouble knowing which foot to pivot on when we're making those fast turns.' And the kid said, 'That's right, Sergeant.' So the D.I. lifted up his foot and he slammed it down on the kid's foot and he said, 'Now, pivot on the one that hurts.' I never forgot that.

"Meanwhile, we were just stymied. We couldn't get anywhere and we knew damn well that the word had come down from headquarters: Don't deal with these guys. They're civilians and the Marine Corps takes care of its own and blah-blah-blah, you know, the old baloney. So we kept walking around trying to interview guys and we were really being—the right word is shunned—I mean, guys were civil, they would say, 'No, sir, I'm sorry, I can't help you with that, sir.' We couldn't get to first base. So Berman said to me one night, he had a bottle of bourbon in front him, and he said to me one night, 'We're not getting anywhere with this goddamn case and I'm not going to spend the whole summer down here if we can't at least make a case out of this.' He said, 'We got to do something dramatic. Come with me.' He just got this idea out of the clear blue sky. So the two of us walked out. I said, 'Where we going?' He said, 'The NCO Club.'

"Now, the NCO Club, which was set aside for drill instructors, was a kind of inner sanctum at Parris Island at that time. It was like Skull and Bones at Yale. It's where the drill instructors hung out. You know those guys put in six or seven days a week, oh, my God, doing everything they demanded of the recruits. So this was their relaxation time where they booze it up a little bit late at night. It was about 10:30, 11 o'clock. We were staying in the bachelor officers' quarters, which they turned over to us, the defense team. That's where we slept, where we prepared our stuff. We went over to the club. I said to Berman, 'What are you going to say?' He said, 'I've got to talk to these guys.' I said, 'Boy, it's a dangerous tactic, Zuke.' He said, 'Well, we're at the moment where we've got to do something.'

"He walked in. He was pretty steady on his feet, but not too steady, and he walked in and as we crossed this room it got quieter and quieter. You know it's a raucous place. And pretty soon it was silent. Now you've got to picture, this guy was about six two, gangly, skinny, very Jewish-looking guy, big bent nose, beady eyes. He was like the classic stereotype of a Jew. So he gets up on the table and he stands there, it seemed to me a half-hour, must have been a minute, until the place was deadly silent.

"Then he said, 'You guys don't know me. My name is Zuke Berman and this young man, Mort Janklow, is a court-martial expert. We're both Jew boys and we're both from New York City and we're both civilians. And I came down here to save the Marine Corps and if nobody's going to help me, I'm going home tomorrow.' He took about thirty seconds to say this.

"Then he climbed down and we walked out. He said to me, 'Listen, I'm exhausted. I'm going to go to bed. You stay up all night in your room. Somebody'll show up, I think.' Two o'clock in the morning, there was a rap on the window. I forget the kid's name. He was about nineteen or twenty years old, a first-time drill instructor, tough kid from Alabama, you know they were all from the South, and he said to me, 'Listen, I may get my ass in terrible trouble with you, we're not allowed to speak to you. Promise me you won't tell anybody it's me.' I said, 'We won't tell anybody.'

"He said, 'What do you want to know?' I said, 'I want to know what goes on here. How do you train these guys? Is it unusual to go in Ribbon Creek?' He said, 'I've been in Ribbon Creek eight or ten times.' He said, 'As a matter of fact, in San Diego they march them in the ocean. That's why they're called Marines. They fight in the water.' So he spilled the whole beans to me that night.

"After this drill instructor opened the door, all the drill instructors came pouring in. I got letters from drill instructors, retired, saying, 'Oh, shit, Ribbon Creek, I spent more time in Ribbon Creek than I spent on land,' you know. If we could prove this, we knew we could acquit McKeon. So we went to work. I'm very proud of the strategy of that case, even to this day, almost fifty years later. Our approach was not to be attacking the Marine Corps but to be supporting it. Any other lawyer without Berman's brilliance in trial lawyering and my experience in courts-martial would have attacked the Marine Corps. You know, we'd have gone after the big guys.

"McKeon was charged with manslaughter, cruelty and with oppression resulting in death, and other lesser charges. He had a couple drinks that afternoon but he wasn't the least bit drunk. But they tried to paint him as this drunken, sadistic drill instructor. I interviewed him extensively and what we discovered was that first of all he was not a strong drill sergeant.

"The unit that was assigned to McKeon was a kind of inner-city bunch of wisenheimers who had no idea what they signed up for when they joined the Marine Corps. They just liked the uniforms and the sabers. And he was fighting for control of them all the time. He didn't dominate them. Most of these drill instructors, they met the first bus of recruits and within an hour of the time the kids were on the base they were broken. Scared the shit out of them, broke their spirit and then rebuilt them. Eight weeks later they were Marines.

"But he couldn't do that. He didn't know how to do it. So this group got more and more raucous and one night they were just being impossible. And there was a tradition as we discovered later at Parris Island, first of all marching the guys in Ribbon Creek wasn't a bad idea because that's the way Marines fight—in swamps and stuff like that. But also it was a way of getting their equipment filthy and getting them filthy and they had to be up all night to clean their weapons and all that stuff. So it was a really strong punishment.

"So on a moonless night he takes his platoon down to Ribbon Creek and he marched them and what happened was, there was a tidal stream, and apparently out in the middle of the stream was a sinkhole, troutholes, and one of the guys—they all had full field packs on which were about seventy pounds—one of the guys stepped in that hole and went under and panicked and started screaming. And the whole group—this demonstrated to me why they needed the training—the whole group went crazy. There was panic and fear everywhere and screaming and yelling and poor McKeon couldn't figure out where the hell anybody was. And six kids drowned.

"I wrote and argued all the legal points and Berman conducted the trial. That's how we divided the responsibility. This guy Howard Lester was really support for Berman's litigating side. He was an associate in Berman's office. He was there.

"In the middle of this, we discussed calling Chesty Puller. I don't remember who called him. That's fifty years ago. Puller of course was the greatest hero in the Marine Corps. He had five Navy Crosses. He was living in retirement in Virginia, and he agreed to come down to testify. We asked him to come in uniform and he walked into the courtroom with fruit salad from his waist to the top of his shoulder. I never saw so many ribbons in my life. And he was still ramrod straight. When he walked in, we knew the case was over. And of course he was to say if you really train men to stay alive in combat, you run some risks in training. He was very smart about it. We rested at once. We knew we had the trial won then. The charges they found him guilty of were minor compared to what they were charging him with. We knew McKeon was not going to serve any more time, and he didn't. But he was a broken guy; I think he was a broken guy coming in to this trial.

"I also think McKeon was very unhappy with us. He was very taciturn and he didn't like what we were doing. He thought we were going to get him convicted. We kept saying, 'Listen, yes, he marched them in the swamp. This was an accident. It never should have happened. The way it happened is indicative of why there should be this kind of training. But it should be better supervised. There should have been another drill instructor present.'

"It was all pro bono, as I said. I think we did such a good job defending him that he became convinced he was innocent and therefore he didn't need us. You know what I mean? It disappointed me a little bit that I never had the playback from him. In addition to not expressing the gratitude one would normally have expected in those circumstances, he was not forthcoming with us. I remember when the trial was over I thought he would come over and embrace me and say, Listen, you know, I know what it must be to be a kid with no money to come down here and spend—I ended up spending the whole goddamn summer there frying my ass off. And it set back my law practice. The little firm I was working for wasn't very happy. But he never said a word. I had to be satisfied with my own sense of accomplishment and with Berman's accolades. I loved him. I just thought he was a great character and he was a great lawyer. Berman earned a lot of respect from the case. It was the first time he had big national press.

"You can't imagine what a victory that was for us. Because they wanted to really put him away. He didn't even serve the nine months. All he got was the time he'd served during the trial." (The next month, September of 1956, the Secretary of the Navy reduced the penalty to three months in prison, already served, and reduction in rank to private. The bad conduct discharge was pardoned and McKeon was transferred to Cherry Point as a private.)

"McKeon was a very hard guy to read. He had levels of sensitivity. I don't know what it was. There was no joy in him. His responses were always very contained. He was withdrawn all the time. He had his ribbons on. The mistake that caused all this was he never should have been a drill instructor. Because the drill instructors are of a type— they're physically overwhelming, they're demanding, they're fired up with patriotism. They're real Marines. McKeon never should have been in that job. He wasn't tough enough. You know, those guys, those drill sergeants are murderous guys. I mean they're just really tough.

That's why their troops are so good. They are oppressive in that, I mean, 'Drop down and give me a hundred push-ups.' And then they do it with you. Those guys are some physical specimens.

"Let me put it to you this way: If I were in combat, what I would want next to me in the foxhole is any Marine drill sergeant. Not an Army commando, not a Ranger, a drill sergeant. Those guys, first of all, could go for thirty-six hours without a wink of sleep. They were amazing guys. They worked like dogs for nothing. I mean, they got paid nothing and in that heat. They were up at four in the morning. They didn't go to bed till midnight. And he was just not cut from that cloth.

"The one thing that's so amazing is, Marines will in fact die for each other. Individual units have performed brilliantly in the Army, but there's something about the tradition of the Marines which is much stronger. When you look at the commercials and you see the kid standing there with the saber in front of him, proud to be a Marine, that's very appealing. Everybody hates the training, but once you're through it you're so proud you went through it. And that's what they work on. It's quite brilliant.

"The case is so long ago and yet it's so fresh in my memory. I can still feel the emotions that I felt walking into that goddamn drill instructors' hall. Berman died fairly young, in his middle or late sixties. He lived a hard life: He was a litigator, he drank a lot, he was a hard-drinking guy but never out of commmission. He was always at the top of his game, a great human psychologist and a great actor.

"Although I did not become a literary agent until 1973, when I helped Bill Safire with his first book after he left the White House, I was married during the McKeon trial to a woman who had an uncle who was a movie producer. And he said to me, 'Do you think that would be a good story?' I said, 'I think it would be a great story if you want to try to film it.' He said, 'I wonder if we could get the rights.' I said, 'I don't think this guy's got any money. Let me see if he'd be interested.' So I talked to McKeon about it, and he wasn't very interested. Nothing ever happened with it. I was there till the very end of the trial, and this was long after it was over.

"But, basically, you have a problem, America, and parents, with your children in the armed forces: Either you're going to train them to really be combat soldiers in which case if, God forbid, they get into combat, they have a chance to survive. Or you're going to insist on softy training where they're not in physical condition and they're not technically skilled enough and they're not courageous enough and they haven't worked together enough to know how to survive. That's the choice. And that was what we hammered away. What the Marine Corps did was right. But that trial was a very formative experience for me because I really believed then and believe now that, in the end, you do soldiers a favor by training them properly."

CHAPTER TEN

JAMES WHEELER

Court reporter,
McKeon case

Because everybody knew this was
a common practice, marching guys in the water
at night, and certainly it was known.

James and Jean Wheeler got married five weeks
after they met at Parris Island in 1956. Both were
Marines. James was a court reporter in the McKeon
case. Jean got pregnant not long after they married
and had to retire from the Corps less than a year after
she signed up. *(Photograph courtesy of James Wheeler)*

Part of the work *James Wheeler and others performed in the legal office at Parris Island was processing for discharge homosexuals and pregnant Marines. One morning, after he had been working there for a couple of years, he said, "I walked into the main office and I saw this beautiful Marine sitting there on the bench, waiting to talk to somebody. She's Sicilian, has this jet-black hair. Beautiful. A beautiful girl and a beautiful woman. As I walked by, I said, 'Oh, man, she is really something.' Then I said, 'No, she's either a lesbian or she's pregnant. That's not going to do me any good.' So I went to my office and about fifteen minutes later one of the sergeants brought her in and introduced her and said she was going to be working in the office. And I said, 'Oh, man!'" Five weeks later they were married.*

"I think we had one civilian court reporter. The other four or five were all Marines. It was a pretrial hearing, the equivalent of a civilian grand jury inquiry, where all of the evidence is evaluated to determine if there is enough to bring charges before a court-martial. Usually more is laid out in this proceeding than is covered at the court-martial. I worked it extensively. I don't remember how much Jean was involved.

"I was in the Corps three years, from 1953 to '56. The event occurred April of '56 and they started the inquiry quite soon after. I was on base when it happened. I think we finished in May. It was a very long process. There were hundreds and hundreds of pages of testimony. They called it a board of inquiry. Most of us referred to it as a pretrial.

"The trial really crushed him. I remember seeing him; they assigned him to the chaplain's office for a while because there wasn't much else they could do with him. He pronounced his name McKewan, by the way. I believe he was a highly decorated Marine, a staff sergeant in his first tour on the drill field. He had been a sergeant of a machinegun platoon in Korea.

"For me personally it was a—how could I say it?—a wrenching thing to go through because I enjoyed my years in the Corps, and I was actually thinking about shipping over and going to OCS. But the way the Corps backed away and left him out there swinging really, really changed my feelings toward the Corps. Because everybody knew this was a common practice, marching guys in the water at night, and certainly it was known. If the commanding general didn't know it was going on, I would have been surprised. I thought the Marine Corps really let him down, and that really bothers me a lot, still.

"I met McKeon in the course of the pretrial. We just spoke casually. I think the prevailing sentiment for him among us was sympathy. There was this weird kind of split between us guys and the Corps. It was like they had to have a scapegoat, because those marches had been going on long before that.

"In fact, it seems to me that when I was a boot we marched out into a swamp somewhere, but they made out like it was a reward. We're gonna go out at night and we're going to wade around in the water. You know, if you're good, you're gonna get to do this.

"I think his junior D.I. testifed that he would have done the same thing. I think it was the drinking, though, that did him in. Maybe he could have had better control of the situation when things started to go bad—I don't know exactly how he was able to respond. If he hadn't had a few drinks, maybe he would have been in a better frame of mind.

"The saddest thing about it was they did a lot of soundings of the water depth there and for the most part the water in that area wasn't even over your head. They said there was some kind of dropoffs. I think they called them troutholes. If you stepped in one of those, you'd go in over your head. I think some of these guys panicked and then started grabbing each other. They thought they were actually over their head and going to drown and that's what really caused it.

"Of course, it was high tide, but I still believe from what I've read that if they'd just got out of the hole, if they were able to step out, they would have been all right. But it was dark and they panicked and started grabbing each other. We saw pictures of the corpses when they dragged them in. It was nasty. And we were just overrun with media people. I think *Life* magazine was big at that time, and they were all over it, trying to buy records of the trial from us, offering bribes, anything.

"I did not get to work on the trial itself. We were tired out just from the pretrial. I think the trial record is, like, seven inches thick. We used to measure it every once in a while. Of course, the case almost wrecked the Marine Corps.

"After I enlisted I came down from New York to Yemassee. When I took my test to qualify for the service, I scored really high. I had one year of college at Columbia University. Anyway, they put me in charge of about thirteen recruits, getting them from New York City to Yemassee. I didn't know what in the world was going on. But that was the beginning. And once you got off that train, you knew you were in a different world. They took us down in buses, heads between our legs. We arrived early in the morning.

"I went through boot camp from September '53 to December. After that I went to Camp Geiger, and ended up in clerical. Originally I was supposed to go to Korea, but they were signing the armistice in the summer and fall of '53 so they changed my orders and sent me back to Parris Island. I spent almost my whole tour working in the Depot legal office.

"We only did general courts-martial. We had a company legal office that did special courts. When I first got on the job, we did a whole variety of absent without leave, desertions, but as time went on we began to get more and more malteatment cases involving drill instructors.

"I would say probably 80 pecent of general courts were for maltreatment by D.I.s. The conviction rate was pretty high. I saw maybe twelve or fifteen D.I.s convicted. They were busted, did time and got discharged. The Corps was really tough. It really clamped down, and this was before McKeon. The maltreatment consisted mainly of punching recruits, smacking them around.

"I took a few shots myself when I was a boot. In weapons battalion, they always stressed you never wanted to be found with a live round on your person because there were cases of recruits wounding themselves or being crazy enough to want to shoot a D.I. Anyway, after you fired the weapon, you would hold out your cartridge belt and the weapons instructor would come by and feel your belt. So this one day he was going by, feeling the belt, and all of a sudden he stopped and he said something like, 'God, what the hell is this?' And he pulls out a round and shows it to me. Well, I knew there was no round in there. So he had palmed it to try to make an example out of me, you know, and he said, 'Who you going to kill? You want to kill your D.I.?' All kinds of verbal abuse.

"The other guy got behind me and the guy in front gave me a shot in the gut and I fell back and the other guy pushed me up. They did that three or four times, warned me, chewed me out some more, and that was the end of it. It had to be intended as a lesson for everybody, because I had never been a problem. I didn't mouth off, and I did what I was told, and I think they just wanted to make an example of somebody. But then, of course, he called my D.I. up, and you can imagine the language. Nothing was off limits back in those days. I knew I didn't have a round in there. I said nothing to my D.I. When they're done with you, that's it.

"All in all, though, I have mostly fond memories of Parris Island because that's where my wife, Jean, and I met and got married. She went through in 1955, and was named the outstanding recruit of her platoon. As a reward she got to stay at Parris Island. I don't think their boot camp was as long as the men's. They didn't get sent to Geiger, either.

"She enlisted from Tarrytown, New York, and I was from Ossining. We were only about seven miles apart growing up but we met at Parris Island. There wasn't a lot of money around the house and she decided to enlist in order to get some educational benefits. She had a brother who'd been a tailgunner in the Marine Corps in World War II, and got shot down. He survived but he lost a kidney.

"Jean wasn't in for even a year. We got married at a chapel on the base. Pretty soon after we got pregnant and, back in those days, you couldn't stay in after you had a baby. She joined September 12, 1955, and got out in May of '56. She worked the McKeon case too but only for a little while. She wanted to stay in. She was a Gung Ho Marine, still is. She's really involved in our Marine Corps League detachment.

"After I got out in September of '56 we came back up to New York and lived there a number of years. I was working for IBM. Later in life Jean went back to school and became a psychotherapist. I also went back to school and got a master's in clinical social work and started working with my wife. We have a private practice here in South Carolina. We call it Serenity Counseling. We have seven acres, a big pond, a really nice set-up. We do everything except young children. We do primarily talk therapy, a lot of marriage counseling. Jean does a lot of individual counseling.

"We're real active in Marine Corps League now. It's a good organization. We try to help Marine Corps families, individual Marines. We're very involved with Toys for Tots. Our detachment just got formed a couple years ago. One of the first things Jean did was get a Marine Corps ball in Rock Hill. We're now planning our third one for the fall of 2005. We've raised thousands for toys, and we have a small detachment."

CHAPTER ELEVEN

GENE ALVAREZ

Staff Sergeant,
College professor

Campaign hats were returned to the MarCor
in 1956, just after Ribbon Creek.

Sgt. Gene Alvarez as a "P.I.D.I." (Parris Island
Drill Instructor) in 1957, a year after the Ribbon Creek incident. Alvarez took ten platoons
of seventy or more through recruit training at
Parris Island. A high school dropout, he went
on to become a college professor and to write
three books. *(Photograph from USMC graduation. Courtesy of Gene Alvarez)*

Gene Alvarez was a high school dropout from Florida who joined the Marines right after he turned eighteen, in 1950. He saw action in Korea, came back to Parris Island, and was a senior drill instructor by the age of twenty-one. He had obtained his high school diploma while serving at Camp Lejeune and made a run at college in the mid-fifties. That didn't work. He went back into the Corps, and his second attempt at college began in 1961. This time he made it, ending up with a master's from Ole Miss and a Ph.D. in history from the University of Georgia.

"I was on the drill field in '53, '54 and again from '56 to '59, both before and after Ribbon Creek. I came back there in June of '56 and attended McKeon's trial for one day. The Yellow Footprints were not there then. A lot of stuff came in after that, the warrior breakfast, the Eagle, Globe and Anchor, graduation. It was amazing the imagination of some of those people they have there, especially those drill instructors, when they had much more freedom than they do today.

"But everybody was scared to death some bad story was going to come up and we were alerted to watch out for reporters lurking around the barracks area. Recruits would challenge people while they were on guard duty, people they thought were not supposed to be there. A lot of courts-martial were starting. A lot of drill instructors were being court-martialed. They'd bust 'em down to private and give them brig time.

"I don't know how many folks know they had a Medal of Honor winner locked up with McKeon in the brig there at Parris Island, for punching out an officer. I read his citation, Alford McLaughlin, from Alabama. Man, this guy had a résumé like Rambo. He got it for action in Korea, 1952. If we'd had ten of those in Vietnam, we'd have won the war. This guy was something else. He died some years ago. His citation was unbelievable. I met him briefly and he was just an old country-looking boy with some freckles, not terribly attractive.

"The big dilemma at Parris Island was, a lot of D.I.s thought they were doing their duty by the rather physical way they did things, thumping recruits. And it worked, believe me. Of course it's against the UCMJ [Uniform Code of Military Justice] and everything else to do that, and so a lot of courts-martial were going on. Most of them followed Ribbon Creek. They just got the word out that it's got to stop, but what really happened was, the thumping didn't stop. It just went underground. It didn't stop until the late '70s, when that recruit, McClure, was killed by a pugil stick out at MCRD San Diego.

"Then there was a D.I. who some way or another shot some recruit with an M-16 at Parris Island, shot him in the hand. Those two things happened as the Marine Corps was coming under congressional investigation twice in a period of twenty years, and that's when the abuse really stopped. They'd still have a few instances of it there, but nothing like it used to be. But we don't have the kind of recruits now that we did either. It was expedient for us to do that, thump them.

"You know, I'm a retired college professor, and I have five recruits, including a retired major, who have tracked me down. And we try to get together once every year. So we did some things right on the drill field, too, you know.

"I often wondered how was it you could take seventy to seventy-five people, young men, treat 'em like dogs, so to speak, for most of the time, and they come and they will die for you. And that often fascinated me. And of these five who have tracked me down, three of them have told me, they still call me sergeant, they said, 'You changed my life. You know I went to Parris Island, you grabbed me by the stacking swivel and straightened me out and made a man out of me.' They said, 'I owe you that.' I spent thirty years teaching college and I've never got the rewards teaching college that I had taking recruits through Parris Island. I took ten platoons through that I can document, sixty-eight to seventy-five in every one.

"They still do a good job over there, but things have changed with society overall, you know, and one big difference now is that if a recruit really wants to get out, he can get out. Whereas in our time we took 'em in the gear locker and all of a sudden they decided they didn't want to get out, you know. I'd say we had 1, maybe 2 percent incorrigibles, and these were recruits who just didn't care. You could shame them, you could humiliate them, they had no self-esteem whatsoever, and those were the ones we really had trouble with. But usually a good shove in a wall locker or something would get their attention. A little push here and there and it's amazing what you can accomplish. Call it being Dominant Dog, Alpha Dog.

"I'm seventy-two years old. I retired from teaching college ten years ago but I've brought out three books since I've been retired, two of them on Parris Island, the last one more or less a picture history. Then I did a book on Jack Webb. That was really quite an experience. Jack's third wife, Jackie Loughery, and I have become buddies. She was a Miss USA, and she's not hard to look at either. She had a starring role in Webb's movie, *The D.I.*

"The picture came out in 1957. I know the story very well. It's about this D.I., played by Webb, trying to save a recruit. There's no thumping, it's all very positive. The welfare of the recruit is the primary concern of the movie.

"At first, Jack had in mind doing something based on Ribbon Creek, but the Marine Corps said it didn't want to touch that because it needed some good publicity then, instead of bad. So then James Barrett, the scriptwriter, had done a play about the Marine Corps when he was in school at Penn State called *The Murder of a Sand Flea*. And Webb took that play and modified it and they came up with *The D.I.*

"When George Stevens and all of them came to Parris Island to make tape recordings of drill instructors' voices, my platoon was in the lower squad bay across from where Stevens and his people were and Stevens inadvertently recorded one of my junior drill instructors' voices and when Jack heard the tape he had Cpl. John Brown come out to Hollywood where they were making the movie, and Brown was given a major role in it. When you watch the movie, they're standing by the scuttlebutt, the water fountain, and there's Webb, the recruit, Owens, and my former D.I., who has a real froglike voice. He's also with Monica Lewis in a nightclub scene. Jackie, she plays Annie, she's the lead. That was my introduction to the movie and Jack Webb.

"Little did I ever think that I would write a biography of him, much less become an e-mail buddy of Jackie Loughery. She was the first

Miss USA and she also was Miss New York State. Her first husband was Guy Mitchell, the singer, and she dated Sinatra and all those people. Jackie's a lovely lady.

"One thing I will mention about that movie is that Jack was a perfectionist. He wanted everything on the set to be absolutely correct. So the scenes in the squad bays, all the bunks, Webb rattling the GI can, the D.I. room, those things are all so accurate that a lot of Marines swear that movie was made at Parris Island, which it wasn't. It was mostly shot on a movie lot in California and the platoon was a recently graduated bunch of recruits from MCRD San Diego. I was told it premiered at the Parris Island Lyceum.

"How did I come to write the book? A researcher called to speak with Dr. Stephen Wise, the director of the Parris Island Museum, wanting information on Parris Island and the making of the movie, and Steve told him to get in touch with me. We talked and I told him I'd send him what I had, and the researcher wrote back and asked if I'd be willing to write the book with him. I said I'm an historian, not a journalist. But he said, No, we want to go with you. So I did it, and had a marvelous experience and went out to Hollywood and met a lot of fascinating people, Barbi Benton, Hugh Hefner's girlfriend, and Red Buttons and Jay Leno.

"I was born June 25, 1932. I grew up in Jacksonville, Florida. I was behind in school, making poor grades, and I wasn't getting along well with my family and so I quit after eleventh grade, and enlisted August 12, 1950, right after I turned eighteen. I was just a poor student and for some reason I couldn't get it all together. The Korean war had just started and I was all impressed with the Marine Corps from World War II. I ended up in boot camp with a bunch of Philadelphia hoods and I was kind of glad when it was over, but, yes, the experience certainly impressed me. It has to. But I was never touched by a drill instructor in boot camp in 1950. I kept my mouth shut and did what I was told.

"I was in Korea in 1952 with the First Marine Division. I spent about ten months up on the line, then I came back and got sent to the drill field. By the time I turned twenty-one, in 1954, I was a senior drill instructor.

"As I was getting out this girl I was dating said, Well, why don't you go to college? I said okay and entered the University of Florida on probation—I'd got my GED at Camp Lejeune after boot camp—but I did very poorly because I didn't know how to study. So I quit college and went back in the Marine Corps, right back to Parris Island, from '56 to '59, so I saw PI before McKeon and after McKeon. But when I went back I found I missed books, I missed reading, and I couldn't get enough of learning. In fact, the officers were often amazed to see that a drill instructor had a set of *Encyclopedia Britannica* in his quarters.

"Then I decided, when I couldn't get a commission because of an eye injury, that I wanted to be a college professor, so I got out in 1961 and this time went straight through college. I finished at Jacksonville University, then got my master's at Ole Miss in Oxford, then my doctorate in history at the University of Georgia. Why history? I just always liked it. I had a fifth grade teacher who really influenced me. I just always liked history, and I got hooked on the Civil War.

"I taught mostly in Georgia. I taught at several institutions there and then three years at the University of Maryland. I was in the overseas program so I taught mostly military personnel and their dependents, some nationals, in Asia one year, from Japan to Thailand, and the second year I taught in Europe from London to Athens. I taught in Vietnam also, mostly American history.

"The thing about the Marine Corps is, it's just like the Catholic Church. It adjusts to the times. I'm not Catholic. I don't belong to any church but I do know Catholics used to do a lot of things they don't do anymore, like eating fish on Friday. The church has to survive, same as the Marine Corps. There have been several instances where the Marine Corps was nearly dissolved, starting as far back as the Depression years, but it always survives.

"Many forget that recruits once lived in not only barracks [all of the old wooden barracks are now gone] but in Quonset huts and tents when necessary. In World War II rough wooden Personnel Barracks were used, but were abolished soon after the war. Succinctly, Parris Island always changes. The Depot was once isolated.

"The generals, you know, they have bosses too, and they have to adjust and do things their bosses pass down, but what I would say—

and this comes in with the Ribbon Creek thing—I think that Marines back when and today both, they feel they have a reputation to uphold and, even though the rules may not be to their liking, they adjust as best they can. It's about tradition, commitment to tradition, and the desire to preserve that tradition, even in changing times. There's the old joke that Marines don't like to be referred to as ex-Marines. I heard a joke that there's only one ex-Marine, and that's Lee Harvey Oswald."

Like many drill instructors who saw the changes in the training program before the McKeon court-martial, Alvarez feels the Marines have sacrificed much of the toughness that made them special in his day. "This Incentive Training they have today is kind of hilarious but it's the world we live in now and in the Marine Corps it's called Cover Your Ass. I go over there to Parris Island now and they do a good job and the base is just beautiful but I don't really enjoy it because it's just so different from what I remember. They still do a good job; it's just that times have changed. They throw Bronze Stars around like they're nothing today.

"When I'm over there now I see them doing push-ups. As late as 1956 and even later, D.I.s had much more freedom than they do today. The Motivation Platoon before Ribbon Creek was the drill instructor. He motivated pretty well. Also, among any D.I.s, even those with SOPs, people do things different ways. There is no single mold. But D.I.s used to have so much more freedom. Some abused it, and I suppose all of us old guys did, by today's standards. But many credit the former harsh and physical training to the amazing survival record of Marine POWs during the Korean War. As a 1952–54 and 1956–59 D.I., I have very few apologies to offer.

" 'Thumping' covered a myriad of things, from a push to physical abuse. It is important to recognize that not all D.I.s went around punching recruits in the gut or elsewhere. Some did, but from my experience that was not a general rule. Nor were swamp marches. The primary weapon of the D.I. was his voice. I suspect that it still is. A close face-to-face chewing out usually worked. It should also be known that D.I.s had to build platoon morale and spirit. That could not be done through abject cruelty. This was done in strange and bizarre ways at times that civilians may not understand. Lighting the smoking lamp [being allowed a break to smoke a cigarette] was usually an earned pleasure.

"Locker box drill was usually some kind of a physical drill using the heavy and cumbersome locker boxes. The manual of arms using a locker box was one drill, and I heard of the Locker Box 500. I never saw this but understand it was sort of a race with locker boxes. The drills were not common since they could waste training time and cause injuries. The D.I. always kept the recruit off balance. I found that early in the training the recruit adjusted and one could see, even then, the spirit building take place. Many D.I.s had vivid imaginations.

"It is very important to understand that the goal of the D.I. and the recruit program was to instill discipline and obedience, and esprit de corps. The proof is, this was accomplished. Most recruits throughout their lives have great respect for their D.I.s, so although the program could be pretty rough for some boots, something was done right.

"Remember, I am speaking mostly of my experiences and observations from 1950 to 1959. Times have changed and civilians could never understand the psychology of Marine recruit training, especially where any physical means were used in the past. Overall, the D.I.s were good men and the crème de la crème of the Corps, surely after D.I. School. I went through the demanding school twice. Moreover, being a Parris Island D.I. was one of the best and most rewarding experiences of my life.

"As I said earlier, the real changes came in the late '70s, with the pugil death in San Diego and the recruit getting shot at Parris Island. It was apparently an accident, but also at the time the Marine Corps was kind of in trouble. These guys had come out with this book called *See Parris and Die*, and it was the end-of-Vietnam era, anti-military times. The McClure kid got beaten out there with that pugil stick and they sent him back to Texas, where he died. He should never have been let into the Marine Corps. That kid wasn't right. I am not defending that he died or was beaten or what happened and I'm sure there are two sides to the story, but, anyway, that's when things really changed.

"I knew thumping still went on at Parris Island after Ribbon Creek. We just were more careful about it. They even had an officer in my company who was caught thumping and General Greene was pretty pissed at him about it. There were a lot of courts-martial. A couple of them were countered by lawsuits from parents of drill instructors, stuff like that, so I was kind of happy to get off the drill field after I'd done my two years.

"Another thing that is really different at Parris Island from when I was a D.I. is we really relied on stress. The idea was that if they're going to break under stress, break 'em in boot camp instead of in a combat situation. Boot camp today I think is physically tougher than it used to be. Everything's PT, PT, PT, but it's not as mentally tough and I personally like the old school a little better, with the emphasis on mental stress, harassment, little stupid things that make no sense. I read somewhere in *Marine Corps Gazette* where the Marine Corps now took the attitude that you can't learn under stress and so stress is out. But that's just nonsense. And I have proof of it. Ask any Marine what his Marine Corps serial number was and he'll tell you. I still remember mine. It was 1137622 and I went through Parris Island in 1950. You learn under stress, don't think you don't. I try to keep my mouth shut when I go over there now because things have changed."

The Yellow Footprints

Parris Island . . . or San Diego?

The Yellow Footprints, eighty or more pairs of large footsteps painted on the deck in bright yellow outside the Receiving Barracks at both Parris Island and at the Recruit Depot in San Diego, have become a Marine Corps icon, a recognizable symbol perhaps second only to the Eagle, Globe and Anchor insignia worn by every Marine. The Footprints are laid out in platoon formation, and the drill instructor strongly encourages new recruits to hasten as quickly as possible off the bus and onto a pair of footprints. The odd part is, no one seems quite sure of who thought of them, when they first went down, and where—whether it was South Carolina or California. Opinions differ.

On the Parris Island side, Chuck Taliano, who operates the gift shop for the Depot museum, recalls, "It had to be mid-1964 or later, when

Their origin is uncertain, but every recruit to pass through boot camp at Parris Island or San Diego remembers them well. President John F. Kennedy visited the recruit depot at San Diego in 1963, and a set of footprints made where he stood was later cast in bronze. Kennedy's footprints have been moved a few times since. There are eighty sets of the Footprints outside Receiving at Parris Island, 104 pairs at San Diego. The plaque reads: "PRESIDENT JOHN F. KENNEDY MEMORIAL FOOTPRINTS—President Kennedy stood on these footprints during his 8 June 1963 visit to MCRD—San Diego." *(Photograph by Larry Smith)*

Recruit Receiving was moved from the Headquarters and Service Battalion area to the Second Battalion area, which was a barracks they used for receiving recruits for all three male battalions. I believe women were received there too but I'm told they had a separate receiving area because they were not part of the Recruit Training Regiment at that point.

"When Receiving moved from H&S to the Second Battalion, one of the drill instructors there—and it could have been the chief drill instructor—had the idea to paint yellow footprints on the pavement outside the hatch. They painted them in a column of fours and I'm sure it was twenty, which means you had eighty pairs.

"When the recruits came in on the bus, their first arrival at Parris Island, they were instructed to get off the bus and get on the Yellow Footprints. That would have been their first chance here at PI to be standing at the position of attention. That's how the Footprints were painted on the ground. They were also painted in platoon formation so recruits stood in the proper distance back to chest and shoulder to shoulder, the correct military formation.

"The other purpose was for the drill instructor to know if he was going to have seventy-five recruits or eighty recruits; he could count the number of empty footprints and be able to tell if they were all there, to be sure someone didn't stray off for one reason or another. Prior to the recruits being assigned to their platoon or being picked up by their drill instructors, when they would have to leave the Receiving Barracks, to go to chow, go to medical or the hospital, they would fall out on the Footprints from the barracks main hatch [front door]. So the Yellow Footprints continued to be in use from that point forward. Anybody who has gone through boot camp since that time recognizes them.

"In 1986, the Footprints moved again when Receiving was shifted to Building 6000, currently in use in 2005.

"Given their use and notoriety, the Yellow Footprints have become an icon, second only to the campaign hat for new recruits, and they are now a symbol, an icon for all Marines, irrespective of rank, because of the implied message: 'I stand with the Marines.'

"To digress for just one moment back to the original posting of the

Footprints: Marines are scroungers. So whoever that drill instructor was who dreamt up the idea went to Depot Maintenance to scrounge some paint, and the only paint that they could give him was the paint that they use for center stripes or to mark curbs for no parking. And that's how they became the *yellow footprints*. It could have been any color. That was the free paint. Whoever stood there at a position of attention they traced his feet on a piece of cardboard and then cut 'em out, and that was the template they used. They probably used the boots of a drill instructor

"There's always been contention as to who had them first, whether it was here or San Diego, and I'm not sure that answer is available. What I am told by drill instructors who were here at that time is that the drill instructor here at Recruit Receiving had the idea first but he immediately got on the phone and called his counterpart in San Diego and said, 'Here's what we're going to do, what do you think?' So, in fact, San Diego may have painted theirs first. In my opinion, it's irrelevant who thought of it but both depots had them right about the same time.

"In any case, one thing led to another, and we came up with the idea of putting them on a floor mat for sale as a fund-raiser, and we've sold about 700 of them at forty bucks apiece. The money here benefits the museum and goes to scholarships as well. We've done so well with that we also have ordered pins and clothing. The Footprints are now a registered trademark. They're unique."

Meanwhile, Don Garland, commander of the Yellow Footprints Detachment of the Marine Corps League, had this to say: "The Receiving Barracks at Parris Island has been located in a number of sites dating from the Depot's beginnings in 1915. The first one was situated at the old State of South Carolina quarantine station, where Traditions, the officers' and staff NCO club, stands today. During World War II and Korea, receiving was conducted at Building 145 in the H&S Service Battalion complex. Receiving was moved to Panama Street, Building 631, Second Battalion, in 1964, as Chuck Taliano says."

So the prevailing Parris Island opinion is that the Footprints first appeared with the move to Second Battalion in 1964. However, Donald Cain, a former drill instructor, thinks they appeared as early as

October or November of 1960 and says they were definitely recalled by recruits from July of 1961.

But out in San Diego, Ellen Guillemette of the Recruit Depot museum reported that the Footprints were originally located in the back of Building One. "The first images of them in their current location appear in the grad books from 1963."

This tends to support the point of view of Frank Roller, who served in the Corps thirty years and one month, retiring as a sergeant major in June of 1984. He and his wife now live in San Diego. The actor and former Marine Lee Ermey is a good friend, Roller says, and he sees Iron Mike Mervosh three or four times a year.

Roller joined the Corps in June 1954 out of Buffalo, New York. He was only seventeen, and his mother signed for him because he and his stepfather did not get along. He weighed a hundred and thirty and stood five feet five. "I was still growing," he said. "Eventually I got up to five feet eight."

Roller served as a sergeant instructor at Officer Candidates School at Quantico, Virginia, from 1958 to 1961. He was in Vietnam from 1965 to 1968 as an artillery operations chief. He was assigned to the drill field at the Recruit Depot in San Diego from 1968 to 1971.

Here is what he says about the Yellow Footprints:

"Rumor has it that, back in 1960 or '61, President Kennedy visited the Depot and they put a set of brass plates out there where he stood on the inner side of the Parade Deck of what used to be Receiving Barracks. You had the theater there, and off to the left of that theater as you're facing it was the old Receiving Barracks. So after they put the brass plates in where President Kennedy stood, they put all the other Yellow Footprints down, to make it look uniform.

"Originally they were located in back of the Receiving Barracks. Then we moved 'em, and I used to cuss them Yellow Footprints because you had to shine that brass every morning. Not me, you know, I made the troops shine it. But that's the story that we got.

"I guess they figured out what a good idea it was, because if the newly arriving recruits put their feet on them right, it was a forty-five-degree angle, and you didn't have to walk up and stick your foot between their two feet to separate them. I think it was eighty sets of prints, and they were spaced thirty inches apart, whatever the back-to-chest was. So you started off giving them the lecture, put your thumbs along the seams of your trousers, feet at a forty-five-degree angle, and you didn't have to correct them on that, because they're standing on the Yellow Footprints.

"I don't know if that story is true or false, but that's what I was told."

PART THREE

MARINES AND THE MOVIES

*Image Matters
Greatly*

HOLLYWOOD has long been a great recruiting tool for the Marine Corps, producing as far back as 1918 films like *The Unbeliever*, which, as a by-product, induced young men to join up. In 1926 came *What Price Glory* and, the next year, *Tell It to the Marines*. Nearly everyone who joined the Marine Corps after 1949 saw and admired *Sands of Iwo Jima*, starring John Wayne. Other films featuring the Marine Corps included *The Leatherneck, Battle Cry, Gung Ho, To the Shores of Tripoli*. In more recent years there was *A Few Good Men*, which the Corps did not like especially, even though there's a great scene with the Silent Drill

Team, and Colonel Jessup, portrayed by Jack Nicholson in the movie and Stephen Lang in the Broadway play, gives a rugged, heroic harangue from a military man's view: "You want the truth? You can't handle the truth!"

The actor Jack Webb, as cited by Gene Alvarez in Chapter 11, wanted to make a movie based on the Ribbon Creek incident but was talked out it by the Corps. Instead, in 1957, he made *The D.I.*, about Tech Sgt. Jim Moore, a tough drill instructor, played by Webb, who refuses to give up on Private Owens, a failing, recalcitrant recruit, played by Don Dubbins. A very positive portrayal of the Marine Corps and boot camp, the movie was notable for its employment of one of Parris Island's most dominant creatures, the sand flea. The Dubbins character slaps a flea when he is supposed to be lying still, leading to an outrageous midnight search ordered by the D.I. for the corpse of the flea so that it can undergo a decent burial. When a recruit at last finds a flea and brings it to the drill instructor, the Webb character asks: "What sex is it?" The recruit cannot possibly win. The film was based on a TV play, *The Murder of a Sand Flea*, written by James Lee Barrett, who also wrote the script for *The D.I.* Two former Marines, the well-known actors Lee Marvin and Hugh O'Brian, had roles in the TV drama. The story supposedly takes place on Parris Island, but the movie was shot on the West Coast. A platoon of nineteen Marines is named in the credits.

Here is drill instructor Moore's opening address to the new platoon:

(The scene begins in the dark, with Webb banging the inside of an aluminum garbage can. The platoon does not jump out of the racks fast enough to suit him, even after three tries.)

"Rise and shine! Another day to serve the Corps! Just like a bunch of little girls. Again!" (Repeats process.) "Again!" (Repeats process.) "You people are too slow. If you were that slow in combat, you would be dead!" (Paces.) "Dead! You burr-headed idiots do not appreciate my cheerful good morning. When my back is turned you call me bad names. But I won't hear you call me bad names because if I do I'll go to the brig. But I'll be thinking about you people all the time. Do you hear me? [Platoon responds: 'YES, SIR!'] I can't hear you. ['YES, SIR!'] I am not your mother. I will not wake you up like your mother does. Do you hear me? ['YES, SIR!'] When the lights go on in this squad bay, you clowns are supposed to be at attention. But noooh. Oh, no. When the lights go on, where are ya? You're crawling out of them racks. I've got a hundred-year-old grandma who gets out of her sack faster than you people, doesn't she? ['YES, SIR!'] You better square your head and eyeballs, boy. Get 'em on that bulkhead. How long you been shaving? What's the first word out of that filthy mouth? [Recruit responds: 'SIR?'] How long you been shaving? ['SIR, SINCE I WAS SIXTEEN, SIR.'] Those sideburns come off, boy, you understand? We don't wear sideburns in this Marine Corps. ['YES, SIR!'] You know who wears sideburns, don't you? ['YES, SIR!'] Well, you ain't got no guitar. You get it off, even that peach fuzz. In this Marine Corps, you shave from your ears all the way down to your collar bone, hear? ['YES, SIR!'] The only thing you do not shave is your eyebrows.

"Out on that drill field yesterday you people were miserable! You people ain't even a mob. A mob's got a leader. You clowns are a herd. I ought to get me a sheepdog. I've seen better maneuvering in a Chinese fire drill. But I just want you to know. Today's gonna be different. You're gonna drill, drill, drill! And you're gonna do it right, do you hear me? [Platoon responds: 'YES, SIR!'] You ain't here for no picnic. This ain't no summer camp. But you wish it was a summer camp don't you? ['NO, SIR!'] You wish you could listen to radio, eat pogey bait, and smoke cigarettes, don't you? ['NO, SIR!'] You wish you could go to the movies, take women out, and booze it up at the slop chute, don't you? ['NO, SIR!'] You like this training, don't you? ['YES, SIR!'] All right, tigers, give me a growl! ['GRRROWWWWLLL!']

"Rodriguez, if you were completely surrounded this morning by an enemy force of 500 men, what would you do? [Rodriguez responds: 'KILL, SIR!']"*

Here Webb goes and speaks to an assistant D.I. in a fatigue cover, directing him to take the platoon out for a PT run, and then to breakfast. He tells the recruits that, because they were slow getting out of bed, the "smoking lamp" will not be lit, meaning they will not be allowed to smoke after breakfast. There is no "smoking lamp" in today's Marine Corps.

In Chapter 13 of this book, Col. Mike Malachowsky, former chief of staff at the Recruit Depot at Parris Island, tells how he made the movie required viewing for his D.I.s in the 1970s, to show them how far a drill instructor could go to help a recruit succeed, and to demonstrate how it was possible to motivate a recruit without swearing.

A second dominant film that focused on recruit training, *Full Metal Jacket*, came along in 1987, exactly thirty years after *The D.I.* Produced by Stanley Kubrick without the support of the Marine Corps, *Full Metal Jacket* was similar in theme to Jack Webb's *The D.I.*, although with a marked contrast. The first half of *Full Metal Jacket* focuses on a drill instructor, Gunnery Sergeant Hartman, played

* Excerpt from *The D.I.* granted courtesy of Warner Bros. Entertainment Inc.

by R. Lee Ermey, who was an actual drill instructor at the Marine Corps Recruit Depot in San Diego during the Vietnam war years. Again the setting is Parris Island, although the movie was made in England. The recruit training segment focuses on a difficult recruit, just as in *The D.I.*, but instead of succeeding as a Marine, he goes crazy, smuggles a magazine of M-14 rifle rounds from the firing range, and kills his drill instructor before committing suicide. Ermey's recruit-training scenes are dynamic and brutal. He said he himself wrote most of the drill instructor's dialogue, based on his own experiences in training recruits. The second half of the picture follows the platoon to Vietnam.

The movie opens with Gunnery Sergeant Hartman, like Jack Webb in *The D.I.*, addressing his new charges, but the language is much more explicit in its use of obscenity, bigotry, homosexual imagery, and threat, all related to demands for performance by the recruits. The Ermey character is at once filthy and hilarious.

After informing a black recruit that he will henceforth be known as "Private Snowball," Gunnery Sgt. Hartman encounters and names "Private Joker," played by Matthew Modine, whom he punches in the stomach for mouthing off and then declares, "You had best unfuck yourself or I will unscrew your head and shit down your neck!"

From here, he talks to a recruit from Texas who becomes Private Cowboy, and then he finds an inept, plump individual, Leonard Lawrence, played by Vincent D'Onofrio. He names this recruit Pyle, after Gomer Pyle, the hick TV series character. This scene sets up the conflict that will play out between Gunnery Sgt. Hartman and the recruit until the final, bloody confrontation. It is followed by shots of the platoon doubletiming as Sergeant Hartman calls cadence with a voice-over from Private Joker, intoning: "Parris Island, South Carolina, the United

States Marine Corps Recruit Depot, an eight-week college for the phony tough and the crazy brave."

In a notable scene, the drill instructor sends his platoon to bed with their rifles. He has them recite the Rifleman's Creed, which every Marine learns:

"This is my rifle. There are many like it but this one is mine. My rifle is my best friend. It is my life. I must master it, as I must master my life. Without me, my rifle is useless. Without my rifle, I am useless. I must fire my rifle true. I must shoot straighter than my enemy who is trying to kill me. I must shoot him before he shoots me. . . . I will. Before God I swear this creed. My rifle and myself are the defenders of my country. We are the masters of our enemy. We are the saviors of my life. So be it, until there is no enemy, but peace. Amen."

As the first half of the film nears its end, as the platoon is graduating, Gunnery Sgt. Hartman tells them they are now Marines and some of them will go to Vietnam and not come back, and that, while Marines die, the Corps will live forever and, as a consequence, so will they.

Although *Full Metal Jacket* was made without the sanction of the Marine Corps, Ermey said, not a day goes by that he does not encounter someone who says he joined the Corps "because of that movie."

CHAPTER TWELVE

R. LEE ERMEY

Drill Instructor,
Actor

I argued with Kubrick until I was blue
in the face but he would not change that aspect of it.
I have never known a private to ever be slapped or hit in the face.
He was struck no place other than the solar plexus
and it was always an open backhand.

R. Lee Ermey. After eleven years in the Marines, he
became a bar owner, a technical adviser on films,
and, finally, an actor himself who went on to star in
Full Metal Jacket and many other films. *(Photograph
courtesy of R. Lee Ermey)*

Ermey parlayed a significant role as the drill instructor in Full Metal Jacket *into a full-fledged movie career following his retirement from the Marine Corps. He has appeared in* Mississippi Burning, Toy Story I *and* II, *and the Academy Award–winning* Dead Man Walking. *In addition to taking on one movie role a year and doing his show,* Mail Call, *for the History Channel, Ermey is frequently on the road raising money for Toys for Tots and doing charity projects or speaking at Marine Corps or veterans' events. I finally caught up with him by phone as he was getting ready to depart on another trip.*

"The drill instructor's language in *Full Metal Jacket* was exactly the way it was back then," Ermey said. "In fact, I wrote the majority of Gunnery Sergeant Hartman's dialogue with Stanley Kubrick. This was the way recruits were addressed in that era. If something the drill instructor said was humorous—and of course recruits were not allowed to laugh—it would be interesting to them and they wouldn't want to miss anything. But the drill instructor basically talked to them in an extremely loud tone of voice so that he could be heard by all seventy-five, eighty recruits. Everything said to each private, even though he's the one being chastised, needs to be heard by the all the other privates because there's a message in everything the drill instructor has to say."

In 1987, when the film came out, Ermey said Hartman "was warped, too rough, too harsh and too demanding—but he was real." Speaking in 2005, Ermey elaborated: "Back in those Vietnam war days, in 1965, '66, and '67, you have to understand, Number one: Instead of picking up sixty privates out of Receiving Barracks, drill instructors were picking up eighty; Number two: Instead of the twelve-week training schedule, they they cut us down to eight; and Number three: Instead of having eight to ten days off between platoons, we were picking up a new herd of privates four to six days before we even graduated the old privates. So it was day on, stay on. Basically, the drill instructors were a bit grouchy.

"It was a two-year tour. However, you were damn lucky to get out of there with two years simply because drill instructor replacements had to come from the fleet and the fleet was in a war. We were in a wartime situation, overextended. So you tried to get an NCO or staff NCO away from the division to come to D.I. school and the division would not let him go. They had those individuals in the combat zone. They were short even at that. So it was impossible to get replacements to come to D.I. school.

"We basically stayed down there at MCRD San Diego about thirty months, working pretty much 100-hour weeks. Our biggest problem was now we had eighty privates instead of sixty and we only had three drill instructors and, for a week of that, we're spread between two platoons. The other problem was we had lost four weeks of our training time. So a drill instructor really didn't have time to stand up with a private and make sure he did twenty-five dead push-ups for disciplinary purposes. A D.I. found it a little easier to just walk past and give that private a little slap in the solar plexus.

"We were keenly aware that every private we trained was on his way to Vietnam as soon as we finished with him, and that weighed very heavily on our minds, considering now that we only had eight weeks to train them. They were no longer being filtered into the peacetime Marine Corps. They were headed for war, pure and simple, so it was very important that they be trained properly. And if it meant being sometimes a bit inhumane, then that's what we had to do. We were acutely conscious that some of them would not be coming back. The first thing we would do as soon as we hit our table at the mess hall was open up *Stars and Stripes* and find out if any of our privates were there in the obits.

"The principal objective of recruit training then, as now, consisted of learning discipline and learning to work as a unit rather than individually. Physical fitness was very important too. I would say those three aspects were our key goals: instant obedience, working as a team, and being fit physically. We ran them every morning, three miles before chow. We ran in regular old combat boots, we didn't run in tennis shoes. Reveille was 0445.

"The slap in the face that Private Joker got in *Full Metal Jacket* was not done in the real Marine Corps. I argued with Kubrick until I was blue in the face but he would not change that aspect of it. I have never known a private to ever be slapped or hit in the face. He was struck no place other than the solar plexus and it was always an open backhand.

"Private Joker gets punched in the movie because Kubrick had to have that, for theatrical purposes, and there was no changing his mind. I was technical adviser as well and a technical adviser must be a salesman. He's got to be able to sell the producer-director on his way of doing things. And theatrically a slap in the solar plexus was not adequate for Stanley Kubrick.

"And every talk show I ever went on when I was promoting the movie I made it a point that drill instructors never were allowed to maltreat a private. These privates were not abused, plain and simple. A shot to the solar plexus? Hell, my dad did worse than that to me and I never was injured.

"The Marine Corps has never condoned maltreatment. It was just that, during that era, we were staff NCOs left down there to run the show with no supervision, no officers. We would have series officers, but they were in transit: They would come in and be with you for three weeks until they got their orders to Vietnam and off they'd go. So you had very little supervision. The staff NCOs basically ran the show. The reason they got a little rough with the troops, which was the backhand slap to the solar plexus—basically what it amounted to was a kick in the pants—was because of the time pressures and the numbers. Of course a slap to the solar plexus and the verbal reprimand that followed were so intimidating that the private two years later would be telling his friends how he had received a horrendous beating.

"I had a lot of problems with Stanley Kubrick over the scene at the end of boot camp, when Private Pyle shoots Sergeant Hartman. I argued until I was blue in the face over that, too. Occasionally, in real life, we would find rounds hidden by privates that they would get back from the range. It did happen, but certainly never a full magazine. A round, maybe one round. See, Gustav Hasford, who wrote *The Short-Timers*, on which the script was based, knew nothing about drill instructors. He felt that drill instructors were there to torture privates, that they enjoyed torturing privates, and that there was no reason for anything that was done. As a matter of fact, there was one scene where

Hasford has the drill instructor call the squad leaders in the head and urinate in a commode and they bring Private Pyle in and stuff his head into the commode. I was able to talk Kubrick out of that one.

"But you have to understand that, when you do a movie with directors and producers, you have to be a salesman. You have to talk them out of their evil ways. They go for that drama, boy. For example, I know of no instance in the history of the Marine Corps where a drill instructor was killed by a recruit, yet that was Kubrick's big payoff for the boot camp sequence. Now, Private Pyle was mentally unbalanced, but even at that it had never happened that a private had killed a drill instructor. But it was written in the book and it was in Stanley Kubrick's screenplay and that was the big climax of his boot camp segment. I tried to talk him out of it. I explained there was no way a private was going to get a full magazine of rounds back from the rifle range. They were searched generally three or four times, and every piece of their belongings was gone through. But even at that you still would get back and, somewhere along the line, you'd find a round.

"I never received any negative reaction from the Marine Corps on *Full Metal Jacket*. As a matter of fact, I got nothing but compliments.

That jelly donut. R. Lee Ermey as the fearsome Gunnery Sgt. Hartman brandishes the jelly donut discovered in Private Pyle's footlocker. Someone is going to pay. (Full Metal Jacket © *Warner Bros. Inc. All rights reserved*)

And even today, seventeen years later, there's not a day goes by on the base but what at least one person doesn't come up and tell me I'm the reason they're in the Marine Corps. *Full Metal Jacket* was their motivation. I think that movie, *The D.I.*, and *Sands of Iwo Jima* were probably the most influential pictures of all time as far as recruitment goes.

"I'm a little upset with the Marine Corps right now because of the 'three tattoo' thing. I think that's the most ridiculous crock of crap I've ever heard in my entire life. To get in the Marine Corps now, of course, you must be a high school graduate and you can't have more than three tattoos. And I think they're losing out on a lot of really good kids. I mean, one of the reasons the drill instructors used to be so harsh and intimidating was because of the quality of the privates we had. I mean, who joins the Marine Corps? Tough guys do, like the Hells Angels, for Christ's sake. I had a number of Hells Angels in my platoons. Big-ass damn 220-pound Hells Angels, six foot two, six foot four. I'm six feet tall and I weighed 170 pounds. If you didn't have the tool of intimidation on your side, how were you gonna make this man do what you want?

"Contrary to popular belief, you can't just tell a kid, Okay, you're gonna put this eighty-pound pack on and you're gonna run over here and jump in that mud pit and you're gonna crawl, belly crawl, through it. Unless you can intimidate this private to a certain extent and make him fear you more than he fears the water, or fear you more than he fears the pain, he's not going to do it. I've had a private swear he could only do ten pull-ups, but if you're down there, yelling right in his face and intimidating the hell out of him, and he does that eleventh pull-up, he now knows the rest of his life he can do eleven pull-ups and he'll never have any trouble doing that eleventh pull-up from there on out because he has proven to himself that he can do it. But it's painful and, in order to induce a private to endure a lot of pain such as that, well, he has to be more afraid of you than he is of the pain. That was always the premise of our training.

"I was born March 24, '44. I never graduated from high school. I was in the Marine Corps when I was seventeen years old. My father and mother were more than happy to come down and sign for me. We had moved to Zillah, Washington, from a Kansas farm, and I was a kind of hellraiser. I was in trouble with the law and I was strongly encouraged by the judge to either go in the military or go where the sun never shines, and I ended up going in the Marine Corps.

"I had been joyriding and drinking. There was no drug problems back then. It wasn't drugs, it was joyriding in somebody's parents' car with wine and beer. My dad had been in the Navy, so the first thing I went to try to join was the Navy, then the Coast Guard. They wanted nothing to do with me simply because I had a juvenile record, but the Marine Corps accepted me with open arms. Back in those days the Marine Corps thought, well, these kids, these so-called juvenile delinquents, were just kids that had a little too much of a wild hair. I had many privates who said the judge had given them a choice and they ended up in the Marine Corps and they were great privates. I never had a problem with them.

"I was never trained or stationed at Parris Island but I have been there plenty of times since. Boot camp at San Diego in '61 was twelve weeks plus a week in the Receiving Barracks. They had the Yellow Footprints then, right out behind Receiving, but you have to understand that Receiving today is not where it used to be. Boot camp was very difficult. There was a lot of PT. I screwed up in some way and pissed my drill instructor off and I owed him one million step-ups. So every evening, when letter writing time came about for everybody else, I was up in front of his duty hut doing step-ups.

"There was a step-up board there about knee high and you did step-ups and you sounded off: 'Sir, one, sir; sir, two, sir; sir, three, sir.' You'd go up with one foot, bring the other foot up to the top, then bring the first foot down, then the second, then step up with the second. I could do a couple hundred at a session. He never did get his million, but a lot of my sweat, gallons of it, fell in that indentation where my feet hit the deck, I mean the sand.

"I came out of boot camp as a private. I think we received our MOS before we left the Depot. I went to Second ITR [Infantry Training] for another month at Las Pulgas. My rifle training was where the university is now at Torrey Pines.

"I went to Vietnam after I got off the drill field in 1968. No one volunteered for anything back in those days. You just went where the

Marine Corps said you were going. I was fine with going to Vietnam. I was a Marine. That's what it's all about. I was there about fifteen months. I stayed on. I was single. The only combat I incurred was running from the 122 rockets [North Vietnamese Army rockets] when they came in, which was a couple times a month. I was with Marine Wing Support Group 17. I was in charge of IRO and training, hiring and firing, and taking care of the local Vietnamese workers."

Although a newspaper reported in 1987 that Ermey spent four months in the hospital with shrapnel injuries, he said that was all wrong. "I had a bunker fall on me. I was getting in the bunker and one two twos were coming in. One fell short of the bunker, but the top of the bunker came down on me and dislocated my shoulder. I went down to sickbay and there were so many people down there bleeding and everything I just said, Never mind. I turned around and went back up and worked the shoulder back in. It was just out of joint, but it never did work right after that. This was in DaNang. I was in the Marine Corps eleven years."

Retiring in Okinawa on a medical discharge, Ermey bought a brothel there. "We gutted it and made a bar out of it. It was called SADA, which stood for the Sunday Afternoon Drinking Association. I had the place about two years, and a friend told me the Okinawan FBI was talking to him about me, so I just packed my shit and got out of there. I ended up in Manila, where I managed to meet Ken Metcalf, an American casting director. He still calls me when he comes back to the States." Through Metcalf, Ermey began to work on military movies as an adviser. "I was assistant technical adviser on *Apocalypse Now* and had a small part in it, and I was technical adviser on *The Boys in Company C* and *Purple Heart*, all made in the Philippines. *Full Metal Jacket* was my fourth or fifth film.

"I didn't contact Kubrick. He called me. By that time I was in Clinton, Illinois, building a nuclear power plant. I was the quality assurance engineer there. I had quit the movies. I had decided, Screw this. I had a family by then and I couldn't live on one film a year. So I got a normal job. Meanwhile, Kubrick had looked at all the Vietnam war shows and he said my name kept popping up. He went to Sidney J. Furie, who directed *The Boys in Company C* and *Purple Heart*, and Rick

Natkin, who wrote them. Kubrick found Sidney and then Rick called and asked if he could give Stanley Kubrick my phone number. Then Kubrick called. And that was just too much to pass up, so I quit my job and went to England. I was there foureen months. When I was there I got in a car accident, broke my collarbone, dislocated my hip, and broke six ribs. We were shut down three months waiting for me to recover. Stanley would send a car for me every day, I would hobble out, the car would take me to Stanley's house, and we would rewrite the first half of the film.

"Most of the time I was able to get him to change his mind, but there were a few things, like the killing in the bathroom, that just had to be. I was able to negotiate him out of the other toilet scene. Another thing that I couldn't talk him out of was some kind of army boots that he got from the Danish army or some damn place that looked nothing like jungle boots. I didn't have a problem with the second half of the film. It's all hypothetical stuff. You're in the City of Hue and it's hard to say if something like that would happen."

Vincent D'Onofrio, who plays the beleaguered Private Pyle, went on to a sucessful career, acting in more than fifty films and starring in *Law and Order: Criminal Intent* on television. "Vince has done very well for himself," Ermey said. "I enjoy, I really like his work. And Matthew Modine [who starred in the film as the combat correspondent Private Joker] is a good kid. What you see with Matthew is what you get.

"As for me, I've got a movie coming out next month with Tommy Lee Jones called *Man of the House*. My objective is to try to do about two feature films a year. And, of course, I'm the host of *Mail Call*, which is one of the highest-rated shows on the History Channel. I've done about seventy of those.

"I met my wife, Nila, in Manila, and we have been together twenty-eight years. We have four children, all grown. Right now I spend more time with the Marine Corps than I do with my own family. I've been involved with Toys for Tots for forty years, for the Corps. I take two months, November and December, every year, and I do anywhere from eight to ten Marine Corps birthday balls as a guest of honor or guest speaker. I also do Veterans Day in November, parades and things, and that consumes the entire month. I did three birthday balls in Okinawa last year and five in the states, including one on the East Coast and one on the West. I did the New York City Police Department Birthday Ball last year.

"I'm also involved with a couple of charities, the VFW and the Unmet Needs Program. We look after families of the troops fighting the war. If somebody loses a family member over there and they come to us, we'll help. We're at *unmetneeds.com.* So I do a lot of personal appearances. And I still have old privates popping out of the woodwork wanting me to sign their yearbooks. I think I trained about ten platoons, eighty to a platoon.

"What makes the Marine Corps special? The drill instructors. They are still of the same mold as they were back in my time. They are tremendous role models, and their leadership is second to none. These young people admire their drill instructors. They may fear them slightly until the last few weeks of training but after you get into, say Week 10, they kinda see through ya, they kinda see through the act.

"Let's face it: Drill instructors were not shit out on a rock, and the sun didn't hatch 'em. They graduate from the D.I. school and of course the D.I. school only takes the most motivated and squared-away Marines, so the Marine Corps drill instructor is the best the Corps has to offer. The leadership and role model qualities are second to none.

"Nobody ever forgets his drill instructor. My drill instructors were Sponnenburg, Dvorak, and Freestone. Freestone and Sponnenburg are since gone but Dvorak is alive and well and in Nebraska. And if I ever need motivation, that's who I call. He's still there for me.

"These guys who come into the Marine Corps want the challenge. Many of them are beat down by their parents and are totally convinced by the time they're seventeen, eighteen years old that they're useless, worthless, that they'll never accomplish anything, they'll never amount to anything. The drill instructor's job is to turn that around, and that's exactly what he does. A compliment to a private from a drill instructor back in my time was, 'You're not as fucked up as I thought you were.' I mean, you could just see the private's chest swell up. Of course, he was unsure whether that was a compliment or not, but it sounded like a compliment."

CHAPTER THIRTEEN

MIKE MALACHOWSKY

Chief of Staff, Parris Island
June 2003–July 2004

We always have brought everybody
in the middle of the night. —Sgt. Major Rick Arndt

It's for psychological reasons.
They start the yelling at the airport.
—Col. Mike Malachowsky

Colonel Mike Malachowsky, left, with Sgt. Major
Rick Arndt, at the Recruit Depot at Parris Island in
2004. Colonel Malachowsky, the chief of staff, was
reassigned to the U.S. Special Operations Command
in July 2004. *(Photograph by Larry Smith)*

This interview began in the office of Rick Arndt, who was sergeant major of the Marine Corps Recruit Depot in Parris Island at the time, which was the spring of 2004. Colonel Malachowsky, second in command at the Depot, joined us shortly after, and the discussion ranged over a variety of subjects relating to recruit training.

"I used to tell my D.I.s that one staff sergeant almost put the Marine Corps out of business in 1956, I mean literally put us out of existence," Colonel Malachowsky said. "Several things saved us: Number one, we had a great propaganda machine. And we were right up front like we've always been with stuff. Right up front. We said we fucked up, we got to fix it. The other thing was, contrary to the way it is today, we had the highest number of veterans in Congress. We had just come out of World War II and Korea, so our reputation was exceptional. But we don't have that now. When I was in high school I knew all the teachers had been in the miltary, a ton of 'em. Three months ago I gave briefing to 130 educators, and when I asked how many of them were veterans, not one raised a hand. I mean, we're doing exceptionally well in Iraq, but as far as Congress goes, it's a different story. When I was the commanding officer at the Recruit Training Regiment here at Parris Island from June of 1999 to July of 2001, we'd get these congressional delegations, and I'd always get the question, 'Why do we need the Marine Corps? We don't need the Marine Corps. The Army does that.'

"The only two things I made mandatory at D.I. school was they had to read *The U.S. Marine Corps in Crisis: Ribbon Creek and Recruit Training*, by Keith Fleming, and they had to watch the Jack Webb movie *The D.I.*

"I never found out if Webb's movie, which came out in 1957, was a coincidence or was made in response to Ribbon Creek, but there was one main reason I made it mandatory viewing for D.I. school students at Parris Island and two subsidiary reasons: The main reason was that the character Jack Webb played, Sergeant Whatever [Jim Moore], considers it a personal insult if he cannot take a recruit and make him a Marine.

"Why is that important? I felt that way when I was first on the drill field, in San Diego as a series officer, from January 1978 to January of 1981, and I still had this same attitude twenty-odd years later when I took over the regiment at Parris Island. I discovered that drill instructors—more so in the '70s—were looking more for ways to get rid of kids than they were trying to work with them. Their attitude was anathema to what Jack Webb's attitude was in the movie: I'm going to make them Marines, whatever it takes.

"The reason that attitude permeated the '70s was because D.I.s had come under intense scrutiny and evaluation. Everything they did was evaluated in fitness reports and their instructor evaluations were all based on statistics. It was ruthless and cutthroat in those days, with a lot of backstabbing going on, with drill instructors trying to outdo each other, one-upmanship, stuff like that. Let's say, for instance, you got a platoon of 100 recruits and 99 of those recruits scored the max on the physical fitness test, the PFT, but you got one recruit scoring 200 instead of the 300. That 200-point guy is bringing down your platoon average, so you find a way to get rid of him. Normally, in those days, we had the correctional custody platoon and, within the parameters of the SOP, you'd just stay on that kid, stay on that kid, stay on that kid, and finally he would either do or say something that was interpreted as disrespectful, and he was run up for nonjudicial punishment—sent to correctional custody for seven days—and he became somebody else's problem.

"So you found a way to get rid of him, but it was also a vicious circle, because you're wiping your hands and saying, Oh, boy, I got rid of that guy, but maybe the week before that somebody else from another series did the same thing. So you get back to your platoon, and here's a guy that just got out of correctional custody after seven days reporting to your platoon and he's running a PFT of 150. The attitude was, I'm going to find a way to get rid of these 'weak guys.' They were making the grade, but they were pulling the percentages down.

"So what we needed was an institutional attitude change. I used to tell my drill instructors, I said, 'You know, if a kid shows up down here and he's already doing seventeen or eighteen pull-ups and he is maxing sit-ups and runs like the wind, you know Bonzo the Chimp can train that kid. Where you need to focus on is those kids who are in the bottom third of that platoon. That's where you earn your pay as a drill instructor, taking those kids and making them Marines.'

"And that's why the movie was important. The whole movie revolves around this one recruit who's 'the problem child' who causes the platoon problems. The Jack Webb character goes and argues to keep him, when the company commander wants to drop the kid. In the '70s, this attitude was unheard of. If there was a reason to drop somebody, whatever, the slightest, Boom! You got rid of him, especially if he was a problem child.

"So we needed an attitude shift and that's what I wanted to get through to the new drill instructors: 'The only thing I'm asking is that you have that kind of attitude.' Now obviously, not everybody can be a Marine, so if you have tried everything in your bag of tricks and the kid still ain't getting it, but you can look yourself in the mirror in the morning and say, 'I have given my all, and this kid just ain't getting it,' then that's fine. I can live with that. But I want you to use everything in your bag of tricks and I want you to take it as personal insult if you can't make this kid a Marine. That was the major learning point.

"And these were my subsidiary points, and this has to do with the professionalism, the attitude, and the example that I wanted my drill instructors to set with the recruits. The first thing was, I am extremely anti–recruit abuse. I was enlisted in the Air Force and I got smacked around. This was in 1968 at Amarillo. I had a buck knife taken to my throat because I'd missed a hair shaving. We called them Training Instructors, T.I.s, and this T.I. was inspecting, I'll never forget, we had the old sateens with the white tee-shirts and as he was inspecting he had this buck knife he was kind of sort of cleaning his fingernails with as he walked around. He looks at me and he stops and he goes, 'What are you, some kind of fucking hippie?' 'Sir, no, sir!' 'You on strike against Gillette?' 'Sir, no, sir!' I guess I'd missed a little hair on my neck, so he took the buck knife and just started rubbing the blade up and down my throat. After a couple of strokes I started to feel the blood trickle down and he just yelled at me about shaving closer next time. And, when the inspection was done and I went back inside, I had a nice little cut and some bloodstains on my tee-shirt. Stuff like that was not uncommon, making us duckwalk and other weird stuff. I think it permeated all the services. That was the accepted norm, I guess.

"Why was I in the Air Force? Myself and my best friend had gone down to join the Marines early in 1968, when the Tet Offensive was going on. I had graduated from high school in 1967 when I was sixteen, so I was seventeen in '68 and would not turn eighteen until that July. When I went home and told my dad I wanted to join the Marines, he wouldn't have any of it. He was in the Army during World War II and he said I won't sign for you to go in the Army or the Marine Corps but he finally relented and said he would sign for the Air Force or the Navy. I figured if I'm in the Navy I'm gonna be on a boat and won't be anywhere near the war, and I'm thinking, Jeez, If I don't get in soon the Marines are going to win the war and I'm not going to get to go. Gosh, I'm not going to get to see this wonderful thing called war. Of course, we all find out in later life it ain't as wonderful as we see in the movies and stuff. So what happened was I joined the Air Force, a four-year commitment, and when I got out I worked for a year as a civilian, then went to college at Oklahoma City University, graduated summa cum laude in 1976, got commissioned through the Platoon Leaders Class, PLC, and off I went to the Marine Corps.

"I later discovered the recruit abuse thing really started in the Marine Corps in the '50s. I found some of the old guys came through as far back as the 1930s and '40s and I'd ask them, 'Did your drill instructors hit you?' And they would tell you, 'Absolutely not, for two reasons: Number one, we wouldn't have stood for it, and, number two, the country wouldn't have stood for it.' When I delved into this a little further, I found this out: The Marine Corps got way smaller when we stood down after World War II, from 485,000 to 75,000, and all of sudden, in 1950, boom, we're in Korea, and now we're shorthanded. So they take an honor graduate from a platoon and turn him around. The kid graduates on Friday and they make him a drill instructor on Monday. Well, he didn't have the maturity, he didn't have the leadership, he didn't have the skills. So what happens if you get flustered and you're a young kid and somebody doesn't do what you tell him what to do? It's 'Whack! Now do you understand?' And that's how it happened and then it just propagated.

"I think anybody is against abuse today. You don't need to slap a kid around to make him a better Marine. You don't need to slap a kid around to make him a better anything. You have enough tools in your possession, whether you make them do push-ups or jumping jacks or whatever, to discipline him. You don't need to be smacking him around. The point of order is, at no time in that movie did Jack Webb or any of the D.I.s ever lay a hand on the kid. Now think about that: This is 1957. I don't know if you've seen *Full Metal Jacket*, but the Vietnam-era drill instructor, played by R. Lee Ermey, who had been a D.I. in real life, smacks kids around, punches them. Those kinds of things were in fact going on then but here it is 1957 and you would think that if Jack Webb had punched the kid, like in *Full Metal Jacket* thirty years later, it would have been accepted. But he didn't."

For Malachowsky, recruit abuse has a personal aspect that makes it all the more deplorable: "How would you like it if it was your son or daughter coming here and being smacked around? I don't think you'd appreciate it as a parent. The mothers and fathers of this nation entrust to us the care and cleaning of their most priceless possession, their sons and daughters, and we owe it to them, to the American people, to train these kids to the best of our ability. We don't need to be smacking them around, we don't need to be calling them names. I wouldn't want my daughter to be smacked around, just for the sake of being smacked around. I mean, I was smacked around. Does it make you a better anything? No, all it does is piss you off. We can be extremely disciplined, we can be extremely forceful, we can teach through example.

"Even to this day I used to ask the recruits when I'd get them in a group, usually at the end of training, I'd ask, 'How many of you expected to be smacked around when you got down here?' Invariably, 75 percent of them raised their hands. They get that from the movies. There's not a lot of good Marine Corps boot camp movies out there. Everybody probably has seen *Full Metal Jacket*, and that is what they still expect it to be like.

"The other subsidiary point is, and once again you can argue the artificiality of this, given that the movie *The D.I.* was made in 1957, neither Jack Webb nor any of the other drill instructors uses profanity. What's important about that? Well, he never smacked a kid and he didn't use profanity, but there was absolutely no doubt he was in charge, he was the guy running that platoon. Today these kids, with the music and movies, they hear fuck and shit and I said, If you are chewing out a kid and you're going, 'You stupid fucking bastard,' all the kid is going to hear is blah, blah, blah. However, if you can chew him out without the swear words, it will have real impact.

"When I came in in '68, we didn't have the nasty language everywhere we turned. So when my drill instructor said, 'Fuck!' it had some shock value. But today it's just the opposite.

"Why is that important? Let's face it, those kids are going to leave here with a certain frame of reference. That frame of reference has got to be impeccable. It has got to be lily white clean. Because they are going get out in the Fleet and come in contact with some bad Marines. They're going see some Marines who are overweight or who don't do their job as hard as others. So they need that basic foundation, that framework where they can always refer back to: 'Gosh, you know, the example my drill instructor set, that's what I want to be like. These guys were the best of the best. They looked sharp, they were professional, they treated us with respect. It was difficult, it was hard, but you know what? They were Marines, and they were great at what they did.'

"Even so, in *Full Metal Jacket*, right up to the suicide scene, everything in recruit training at that period of time was fairly accurate. From that point on, the rest of it, after they got to Vietnam, was crap. The Marine Corps hasn't fared well in the movies recently. In *Platoon*, they tried to stuff everything that was bad about Vietnam into one unit; the fragging, the killings, the racial thing, the drugs. Some of the old movies, like *Gung Ho*, with Randolph Scott, and *Sands of Iwo Jima* were good ones, but even *Heartbreak Ridge* was very artificial. I will say the uniforms were worn correctly, the decorations. I always look for that sort of stuff, to see if there's something bad, like wearing ribbons wrong, or rank and insignia wrong, hair wrong. That picture *Coming Home* with Jon Voight was a piece of crap. I thought *Top Gun* did well for the Navy.

"*A Few Good Men* was a good movie, entertaining, but I don't think all that crap happens. Yeah, there are blanket parties but the convolution of trying to hide this thing . . . obviously they had to play into the story. I do love Jack Nicholson's [Stephen Lang created the role in the Broadway play] speech. I've got it written down someplace. Worth its weight in gold. But I don't think a colonel would lose his cool like that and blow up on the witness stand. That was for dramatic effect. But was it plausible? Aw, sure, I guess. But in this day and age, if you got a kid with a problem, you really do look to take care of him. And if the kid's a weak sister, you usually try to bring him along, and, if you can't, then you get rid of him. Once again, you don't have to beat the crap out of him. And you sure as shit don't need to kill him, just to prove a point."

The Crucible

Colonel Malachowsky was one of the officers involved at the very beginning of the Crucible. "Recruits began going to the Crucible about 1996," he said. "It was set in motion by the Commandant at that time, General Charles Krulak. Brigadier General Steve Cheney was one of the developers, and he and I discussed it, so I was privy to the thinking behind it. When I first read about it in the *Marine Times*, I thought this is just another gimmick. You read the article and thought this is what we've been doing all the time anyway. What the Commandant wanted was a culminating event and he took a look at all the other tough schools, the Rangers, the airborne Pathfinder school, stole a little from the Seals, a little from the Rangers. Then we added our own spin, and it became the Crucible. I wasn't a believer until I first came down here and saw it in action. Now all the other services have adopted some form of the Crucible. The AF has Warrior Week. I don't know what the Army calls it, probably, Let's Hold Hands and Go to the PX [the Post Exchange, where you buy things]. The Navy has Battle Stations.

"But, once again, all we did was take a bunch of stuff we had been doing and give it a name. The problem I faced coming into Parris Island as regimental commander in June of 1999 was a high rate of attrition—17 percent for males and 25 percent for females—growing out of the fact that the Crucible, which comes in the eleventh week of training, looked so far away when the recruits first got here. Because when they first talked with the recruiter, all the recruiter talked about was the Crucible. We know statistically that, if a kid is going to refuse to train, to realize that he or she may have made a mistake, it'll usually be around the seventh or eighth day they're here.

"Now, all they think is, 'I've got ten more fucking weeks to do because that's the Crucible,' and that seems like forever. So that was one of the problems we were facing. When kids decided to quit, the long-range thinking was a major factor: 'I've got ten more weeks before I even get to this thing? Screw this shit.' So one of the things I brought back was Phasing. When I was a series officer back in San Diego, we had First, Second, Third Phase in training. It had been that way for years. By this time, Krulak had retired, so I asked, 'Why don't we have Phasing any more?'

"The answer was General Krulak took away Phasing because he wanted 'a seamless transition' through recruit training. I don't know what the fuck that means. What I know is, as an infantry guy, we got a final objective: graduation. And right now we got no intermediate objectives to get these kids through these stages. It's a psychological thing, and that's why my boss, General Cheney, who was the commanding general at Parris Island, let me reinstitute Phasing.

"And it's still in place. On face value, it's semantics. Used to be, when a recruit went through in San Diego, there was a kind of checkpoint for each Phase. At the end of the first Phase, you got to undo the top button of your blouse. The idea was, when little Johnny or little Susie comes to you at the end of the third week in training camp and says, 'I don't wanna be here any more,' one of the tools you have is, 'Look, kid, okay, two weeks is all I'm asking of you. You're going to get to swim, you're going to get to shoot, you're going to get to rappel, all those things you came in the Marine Corps to do.' So it's not, Give me ten more, it's, Give me two more. There's intermediate objectives, something to look forward to.

"We also brought back range flags, which were ended because somebody didn't like them. I said, Well, I like them and I want them back. It's just something the platoon gets to put on their guidon [a pole that holds the platoon's flag] when they go out to the rifle range or someplace as a unit, and it's unique to each platoon. You usually have an artist in the unit, some of these kids are pretty damn talented, and, so long as there's no profanity, no naked women or politically incorrect stuff, it's a motivator.

"Another big step we took was establishing linkage between recruiting and recruit training. We're the the only service that does that. Every recruit who comes to Parris Island has his recruiter's business card, and the recruiter does not get credit for that kid he enlisted unless the kid graduates from boot camp."

At this point Parris Island Recruit Depot Sgt. Major Rick Arndt noted that the drill instructors and the recruiting community did not talk to one another during the 1970s and before.

Sgt. Major Arndt: "When I was a D.I., I never talked to a recruiter—never one time. They put them on the bus and got them to Parris Island and then the individuals here took over. Now there is more communication than ever. The recruits run into social shock along with the rigors of training, which they will get through just by the nature of the program. But the social shock weighs on your mind, so there has to be some link to home and family. A simple phone call from a senior drill instructor to a kid's recruiter saying, 'Hey, old Private Beltbuckle, have somebody write him a letter, give him a couple words of encouragement.' The recruiters are required to write three letters per recruit per cycle. From the time the recruit reports here till the time he graduates, he is supposed to receive three letters from his recruiter, and those letters are kept on file. In fact, when I go out to visit recruiters, that's what I ask to look at. I want to see their book of letters.

"We also get feedback from recruits, who are asked if they received letters by Phase, and we keep track to see if recruiters are writing, because it does in fact help. It helps a kid who's down here, he's scared, he's lonely, he never had to work to be accepted to a team. He's beat, his mind is tired, he's hungry and he's not sure he's made the right damn decision. All these things are going through his mind and he misses his loved ones. So something as simple as a letter from a recruiter is a damn good thing."

Colonel Malachowsky: "When I got here in June '99, we were concerned about attrition. Contact with the recruiter gives the D.I. one more tool if he's having problems with a kid. We used to have the D.I. put the kid on the phone with the recruiter, because if the recruiter did the job right, help might be as simple a phone call saying, 'Hey, Private Beltbuckle, remember we talked about this?' And that'll get that kid through training. So we got that linkage, whereas in the Army it's like, 'What are you calling me about that kid for? He's your problem.'"

Sgt. Major Arndt: "Recruit graduation and pinning on the insignia the day before all started with the Crucible. Used to be you put your Eagle, Globe and Anchor on your uniform and the first time you wore it was visitors' day, which was the day before graduation. I didn't get any visitors. Nobody came to graduation when I went through boot camp. To be quite honest with you, the only thing I was concerned about was getting on the bus so I could go get something to eat. It was nothing like it is now. We didn't have anywhere near the amount of visitors we have now. I don't even remember people being in the bleachers, whereas now they're packed with family and friends. I don't think we even graduated on the main Parade Deck. We thought we had 90,000 visitors last year. We actually had 115,000 registered, and they're watching to see what you do. So what you do has to be correct. It has to be right. It doesn't have to be simple or easy but it's got to be right.

"What General Krulak wanted in the Crucible was a culminating event. I don't think it's a culminating event. A Marine may get emotional when he receives the emblem during the ceremony on a Thursday afternoon. That's important to him but it's no different than when he's dismissed on graduation day. We made the emblem ceremony important, so it's important to the recruit. The important thing to that recruit is that he graduates on Friday, that he has a good reason for leaving Parris Island, a noble reason: i.e., he's earned the title of United States Marine.

"After Parris Island—or MCRD San Diego for that matter—everybody goes through Marine Combat Training. It's just twenty-four days and that's just to give all the other MOS people—cooks, bakers, and candlestick makers—just a little taste of what it is to be a rifleman. Not an infantryman; a rifleman. Then they branch off to go to a school for their MOS. The infantry guys stay on at Camp Geiger, which is part of Camp Lejeune, or, if they're West Coast, they go to Pendleton for SOI, which is School of Infantry.

"Everything they wanted to gain out of the Crucible is what drill instructors have been doing since my first tour in '81. That's what we did, instilled discipline, we pushed people to the next rung on the ladder, tried to make them understand they can succeed and they can push themselves farther than they realized. Everything we did in the training cycle was to teach them to establish the mindset that they can push themselves beyond normal levels and that there is no limit for them."

The colonel and the sergeant major went on to discuss inconsistency in the selection of officer command as opposed to enlisted in the operation of the Recruit Depot at Parris Island.

Sgt. Major Arndt: "To be a sergeant major down here, you have to have been a drill instructor previously. But a commander can come down here, a company commander, a battalion commander, a regimental commander, can come here with no recruit training experience whatsoever, but because he is the regimental commander, he has the billet to execute and affect change in recruit training.

"Colonel Malachowsky is the way it should be: He was a series commander in San Diego, he knows what is going on in recruit training, so when he comes back later on, he knows what he's doing. We both worked at San Diego and Parris Island and they are supposed to mirror one another. Parris Island, on paper and as far as Headquarters Marine Corps is concerned, is supposed to be the lead Depot, meaning any changes in recruit training are supposed to start here.

"They bring people in. An individual has to have a command, but there's no place in his MOS to give him a command, so they bring him down to recruit training, and he'll be placed in command. Pretty soon you have a command structure with no recruit training experience whatsoever."

Colonel Malachowsky: "You'd never think of sending a sergeant major down here or a first sergeant who has not been on the drill field. In the summer of 2004, four of the four battalion commanders are gonna be lawyers, because somebody decided, 'Boy, we need to get these folks experience.' But they knew when they became lawyers that they were not going to follow a command. Why are we starting this here? We should have former drill field experience as battalion commanders. The regimental commander right now does not have drill field experience. Neither does his XO [Executive Officer]. I'm not saying they're bad people or bad Marines. It's a different environment here. I'm not trying to blow my own whistle, but a lot of things I used in my little bag of tricks were based on what I learned as a series commander. We had lawyers and we had administrative MOSs inked to come to the drill field. It makes no sense. Why'd we do that? It is a sore subject—how command selection is done. There are some very strong opinions about that. You wouldn't put a first sergeant or a sergeant major down here without his or her being a drill instructor first. It should be the same thing for the battalion commander and the regimental commander and that's my piece of the soap box. I talked to Major General [James] Jones and he said it was going to change but it's going to take two more years."

After leaving Parris Island in the fall of 2004, Colonel Malachowsky joined the United States Special Operations Command in Tampa. "We basically run all special operations for the United States," he said then. "We have the lead in fighting the global war on terrorism. You can kinda sorta say I'm out of the Marine Corps now, since I'm on the outside looking in."

Colonel Malachowsky was planning to retire in July 1, 2006, and remain in the Tampa area with his wife. "I'm extremely fortunate in that not too many folks get their last assignment where they actually want to retire." As for his career in the Marine Corps, he said, "Oh, gosh, I'd do it all over again in a heartbeat. I'm envious of these young kids who are in the desert right now and doing great things. I'm glad in one respect, in that I know I had a small part in their initial entry training in the Marine Corps as CO of the Recruit Training Regiment. That's extremely rewarding, really important to me. But to see all this crap about, Oh, you know, Today's generation ain't got no responsibility, all that crap, and every day you pick up the paper and you see kids, not just in the Marine Corps but in all the services, doing great things over there. It's a tough war, and these kids are doing magnificent things day after day after day. And so we're doing something right."

PART FOUR

MONTFORD POINT

How Blacks Made It into the Marines

In affirming the policy of full participation in the defense program by all persons regardless of color, race, creed, or national origin, and directing certain action in furtherance of said policy . . . all departments of the government, including the Armed Forces, shall lead the way in erasing discrimination over color or race.

—President Franklin D. Roosevelt,
Executive Order 8802, June 25, 1941

UNTIL FRANKLIN ROOSEVELT issued this order, which came five months before the Japanese attack on Pearl Harbor, no black person had ever been a United States Marine. The American military was not to be integrated until 1948, well after the war, when President Truman issued his famous order to do so. Prior to this, however, there was a great deal of agitation to allow blacks into military service. Roosevelt's 1941 order allowed blacks to join the Marine Corps, as well as other branches of the military. Roosevelt's wife, Eleanor, was a vocal proponent of black participation, as was the Urban League. Another leading spokesman was the railway union leader A. Philip Randolph, president of the Brotherhood of Sleeping Car Porters.

In August of 1942, more than a year after the presidential order was issued, the Marine Corps established a segregated training camp at a former Civilian Conservation Corps (CCC) site that was part of New River, later renamed Camp Lejeune, in North Carolina. This was August 1942. In the beginning, all the black troops were volunteers, trained by white noncoms. As time went on and the troops gained experience, black noncoms took over recruit training. Two of the most famous were Gilbert "Hashmark" Johnson, for whom the camp is named today, and Edgar Huff. These men went on to distinguished careers in the Marine Corps. In all, about 20,000 blacks trained at Montford Point during World War II. More than 565 black recruit platoons, 23,000 men, graduated from Montford Point before it shut down as a training base in 1948.

CHAPTER FOURTEEN

GENE DOUGHTY

Corporal, Iwo Jima

Guess who they chose to be drill instructors?
Guys from the Deep South. I did not know of one white D.I.
from the North. They came from Alabama, Georgia, Florida, and
Mississippi. They were irritated when they found out some of the Northern
blacks were much smarter than they were. They swore at us but I
will say they never used the N word. If they did that, they
were told, they would be incarcerated, sent to Navy
prison, and be given bad conduct discharges.
All in all, they were fair and square.

Gene Doughty. He was eighty-one in 2005, still active
with the Marine Corps Scholarship Foundation. A
Montford Point Marine, he turned twenty-one during the battle for Iwo Jima. In his hands is a Montford
Point Marine Association gold medal awarded for long-
time service to the association. (*Photograph by Ben Chase*)

I first encountered Gene Doughty at a sixtieth reunion of Iwo veterans in February 2005 in a hotel in Fairfax, Virginia. He was standing at the microphone in the middle of a large ballroom, talking about the role blacks had played in the battle of Iwo Jima. Surprisingly, it was the first I'd heard of blacks being involved in the fight. I spoke to him afterward, and we later had a lengthy conversation on the phone.

"I just got through celebrating my eighty-first birthday. Guess where I was on my twenty-first? I spent my twenty-first birthday on the sands of Iwo Jima. I spent thirty-two days there, from D-Day up until it was time to depart for rehabilitation. I was in 36th Depot Company. Three Depot companies of black Marines served there. We were not integrated, of course, and blacks were not supposed to serve in combat. We handled logistics, supplies, and transported ordnance back and forth, but plenty of blacks saw action on Iwo and on other islands, notably Saipan and Bougainville.

"I landed about 0900 the first day. We didn't go in the first wave but we were in the landings that took place on the first day. Only two black companies, the 36th Depot and the 8th Ammunition Company, went in. Their noncoms were white. The first waves went ashore around 0600 or 0700, somewhere around there. I know it was well after nine when I went in. I came ashore in an LST [Landing Ship, Tank] and then switched to an open barge, like a Higgins boat. I was a squad leader. I will never forget how horrible it was. It's etched in my mind like stone. After I got ashore, my job was to try to get to a beachmaster, who would make assignments based on needs, and then try to secure an area he picked out. We had our fifty-pound packs and our rifles, so we were prepared to take on any responsibility. Getting to the beachmaster was a trip in itself. There were several heights of sand, and you had to inch your way. You could hardly see anything. All we saw was sand, and it was all so pliable.

"The Army had a black regimental transport company there comprised of amphibious Dukw [pronounced "duck"—an amphibious transport vehicle] drivers who not only brought in supplies, weapons, and ammo during D-Day but were also tasked with carrying Marines. They had black commissioned officers. Twenty-five percent of them did not make it to shore. It was a horrible great loss. The amtracs, which brought in most of the troops, were tracked, but the Dukws had wheels and once they got ashore they were a whole lot more maneuverable than the amtracs. The Dukws started to come after the first seven waves of amtracs went in. Black Marines had no part in driving the Dukws, although they rode back and forth in them a lot.

"My outfit handled service and supply, whatever came up that was meaningful. We did a number of things over our thirty-two days there. The main depot companies were the ones involved mostly in going back and forth to the ships. These first two black Marine companies ashore, the ammunition company and my Depot company, the 36th, were given specific chores to do, on the island mostly. We were under fire the whole time, and we probably suffered no more than twenty-five casualties, which is not a hell of a lot, but when you put it in perspective, it could have been much greater. Anybody could be hit by artillery fire at any time.

"A Depot company was a service company. Today you would call it logistics. A service company provided help with division stores while an ammunition company delivered ordnance, all kinds, from hand grenades to rifle rounds to machinegun belts to mortar shells, everything. Other duties involving us included security. There were times when several squads were given security duty, standing guard and the like, and I was one fortunate enough to get that job, fortunate because it meant I did not have to do a lot of heavy loading. On March 26, near the end of the campaign, my squad was among those providing security for the U.S. airmen billeted near the airport when we got caught up in that final Japanese banzai attack.

"Although we had several blacks killed going ashore on D-Day, it was a very, very rare situation to have blacks in combat. The March 26 attack consisted of about 300 Japanese and of course you couldn't tell where they were coming from. How they got through there is beyond me. This was like two or three o'clock in the morning, and pitch black. Some of those Japanese were able to infiltrate the bivouac camps. These airmen were from all branches—from the Army, the Navy, the Marine Corps, and a lot of them had their throats slit. They were sleeping, and they had their throats cut. The Japanese didn't want to use any firearms because right away it would become very obvious that their presence was nearby. So they used their knives and swords."

According to the book *Iwo Jima: Legacy of Valor*, by Bill D. Ross, published by Vintage Press in 1985, the attack came at 5:15 at the airfield Motoyama No. 2, where 300 sleeping Marine shore parties, supply troops, Air Corps crewmen, Army anti-aircraft gunners, and Seabees were assaulted on three sides by Japanese armed with automatic weapons, grenades, swords, and knives. The brunt of the attack was borne by the Fifth Pioneer Battalion. First Lt. Harry Martin, 34, of Bucyrus, Ohio, who was to receive the last of twenty-six Medals of Honor awarded for action on Iwo, put up a scrimmage line manned by black troops who beat back two attacks. The black Marines were in the thick of the fight. Privates James M. Whitlock and James Davis were to receive the Bronze Star for "heroic achievement" in the battle. And at least two black Marines, Pfc. Harold Smith and Private Vardell Donaldson, died of injuries, while four others were wounded. In the three-hour battle, forty-four airmen were killed, and eighty-eight were wounded. Nine Marines were killed and thirty-one wounded. Japanese dead totaled 262, with 18 taken prisoner. Lt. Martin was the last Marine killed in action on Iwo Jima.

"My squad and members of the 8th Ammunition Company were in the area, not just our companies, but a lot of people representing other outfits, because this was a service regimental area. A lot of supplies and everything else was there. We made some major contributions in the fight. I was a corporal at that time. I made sergeant shortly after, when we went to Hawaii for rehabilitation.

"We were training for the invasion of the Japanese mainland when the war ended, and my unit took part in the armed occupation of a Navy base in Kyushu, Japan, and also served in Nagasaki and on Guam. I came home to be discharged at Montford Point.

"I was a twenty-year-old kid when I left home and went to the

Marines. This was the farthest I'd ever gone in my life. I didn't know what the South was all about. I had read about the South, read about the racial inequality, all that stuff, but to get there and face it was a different story entirely. You can understand what went through my mind, but thank God I had the determination and the will to make it through. This was a brand-new experience for me.

"Can you imagine fighting for your country and sitting at the back of the bus? We looked at it as fighting on two fronts, the front line at home and the front line abroad. Sure, we showed resentment. We had problems going to and from, going on furloughs. For instance, just outside North Carolina we had some black Marines arrested by local police who thought they were pretending to be Marines. They did not know that blacks were in the Marine Corps. It was the same thing with the rail lines. Going back and forth from New York to Washington was not an easy trip for me.

"I didn't volunteer. I was selected. The Marine Corps came to me. I had finished high school and I had one year of college. They wanted intelligent, educated blacks for these Depot companies and a lot of men came in with college degrees. Some even had Ph.D.s. The Navy had made it very clear that if they had to take black recruits they wanted the best from all over the country. So we're talking about guys—let's say, from the State of Washington—this poor guy couldn't go to San Diego for boot camp. He had to go clear across the country, to Montford Point, Jacksonville, North Carolina, whether he liked it or not. But, outside of the Tuskegee airmen, we probably had the most diversified, best-educated troops in the country.

"Boot camp was pretty much the same as it was on Parris Island, three months long. We had white drill instructors, seasoned salts with many years in the service. A lot of them were just back from the Guadalcanal campaign. In the beginning, we had no blacks trained who could stand as drill instructors.

"Now this will knock you right between the eyes: Guess who they chose to be drill instructors? Guys from the Deep South. I did not know of one white D.I. from the North. They came from Alabama, Georgia, Florida, and Mississippi. They were irritated when they found out some of the Northern blacks were much smarter than they were. They swore at us but I will say they never used the N word. If they did that, they were told, they would be incarcerated, sent to Navy prison, and be given bad conduct discharges. All in all, they were fair and square. You might not hear the same lines from some of my black brothers, but it depended on your attitude.

"The drill instructors did have that Southern charm, and they would be respectful to you in many ways. As I said, they were not abusive, they didn't hit you, but they would get results in other ways, like giving you a toothbrush and test tube containing soap and water and telling you to go clean the bathroom. A toothbrush, test tube, and soap and water, to clean the head. You like that? And getting up at one and two o'clock in the morning and standing at attention in 80- and 90-degree heat, the mosquitoes outpouring. The drill instructor would say, 'You had your supper, let the mosquitoes have theirs.' And the sand fleas were above and beyond. They were horrible. Oh, they were horrible. Horrible.

"Overall, the training was good, somewhat severe, but we came out with a lot of skills that were exactly what the guys from the North needed. I say that very, very cautiously, very carefully, because we were city slickers, as the drill instructors called us, and they didn't like it that we could do things in the streets of New York and Chicago and Detroit that we couldn't do in Birmingham or Florence, South Carolina, or Maple, Georgia. Discrimination was doubly hard on the Northerners," said Doughty. "We also had trouble with our black counterparts from the South. I couldn't understand them, the slang, the ignorance, but we got used to it and we soon learned our black brothers from Alabama and the rest of the Deep South were good men, good people. They just had a different mindset.

"It was very complex. But you had to cope with it. It was easier for those of us who were a little older, a little more mature. The average age of recruits was a little over nineteen, but here were a hell of a lot who were ten and fifteen years beyond, fellas in their late twenties who already had their professions, and some of whom had Ph.D.s. The older ones had a better understanding of what was happening.

"After I got out of boot camp in June of 1943, I went to the schools company. I learned administrative work, wrote up bad conduct discharges and so on. But then as the need developed, I went out into the field. We practiced landings in the waterways outside Montford Point and then again when we went to Hawaii. I was assigned to Fifth Marine Division in November of 1944, when I went straight to Camp Catlin outside Honolulu.

"By the time I came in they had got rid of the '03 Springfield and and we had to learn to fire the M-1. As the Marine Corps will tell you, it pays to make sure you know how to fire the weapon, regardless of whether you're destined to be a cook or a baker or to work with a service company, because Iwo wasn't the only place where black Marines fought. They succeeded very well with Marine combat divisions. In some instances, every black who had an M-1 joined the fray. All you had to do was use common sense and obey your lieutenant. Saipan was a great example. There was a lot of heroism with black Marines there, when they found themselves on the firing line with their white comrades. They even got a Presidential Unit Citation."

Units of the 7th Field Depot, supporting the Fourth Marine Division, were included in the award of the Presidential Unit Citation presented to the Fourth for its action on Saipan and Tinian. The units were the 3d Ammunition and the 18th, 19th, and 20th Depot Companies, according to *Blacks in the Marine Corps*, published by the History and Museums Division of the Marine Corps.

"When President Truman integrated the military in '49, Montford Point was phased out and everyone transitioned to Parris Island. We had a lot of Montford Pointers go on to become career Marines and, after World War II, fight in Korea and Vietnam. Hashmark Johnson and Edgar Huff were among the first five black drill instructors at Montford Point. Hashmark was much older than we were. He had served with the Buffalo soldiers, the Army's black 25th Infantry, chasing Pancho Villa on the Mexican border in the 1920s. He was a mess man in the Navy before he got himself transferred into the Marine Corps and did not get to Montford Point until he was thirty-seven. He became a field sergeant major and later the first sergeant major of the recruit battalion. He was a hard driver, very demanding and determined to see that we recruits measured up.

"Edgar Huff was six feet five, from Alabama, tough as nails and regarded as nothing but the best. He served in all three wars, like Hashmark, and they became brothers-in-law; they married twin sisters. Huff went on to be sergeant major of the Marine Corps. He died several years ago. I was one of his pallbearers. Johnson died in the 1970s at a national convention, on the dais, just fell and died, that was it. They renamed Montford Point for him.

"I served three and a half years, all told. I almost remained in as a career man because we had quite a few nice guys as officers. I was in communication with two of them until just recently—we're talking about a sixty-year relationship. But I got discharged instead, in May of 1946, took advantage of the GI Bill and went back to school. I had three years to go. I got my degree from the City College of New York in physical and health education, and became a recreation youth leader in the Police Athletic League. I didn't remain in that field too long because there was no money in it. I was an investigator for the New York City Department of Social Services for seven years, then got into sales, ending up as service manager in the communications division of Sears, Roebuck. I retired in May 1985. I stayed in New York all that time.

"My wife, Marian, and I got married in 1948. She went to Hunter, so I married a college girl. She went into teaching. We had two children. We just got through celebrating close to fifty-seven years together. We live in the Bronx."

Even through age eighty-two, Doughty has remained active with the Marines. "I've held a lot of different posts with the Montford Point Marine Association, including national president. I'm currently a board member of the Marine Corps Scholarship Foundation. We hold an annual ball every year to raise money. We invite corporate magnates, the Commandant appears, and we have the President's Own Marine Corps Band, and the Drum and Bugle Corps. It's quite a spectacle and it raises a hell of a lot of money. This money goes toward scholarships for sons and daughters of Marines and Navy personnel. Last year [2004] we had close to 235 casualties in Iraq, and we provided scholarships for any one of their sons and daughters, including civilians, who wanted to go to college. So you can see the magnitude of what we have. We did the same thing in Beirut when all those Marines were killed, including civilian workers who lost their lives in that.

"I'm a committee member. Every ball has an annual journal, and it reflects all the scholarship recipients, and there are biographies of our major contributors and other special pages. I work on that. I've been involved in that over twenty years.

"A question that has often been raised to me is, If I had the opportunity to do it over again, would I? And I would say without a doubt, Yes. It brought me a helluva lot of experience, and discipline. Living and dwelling in the streets of New York as a young kid was no easy picnic, and I was lucky to get out of there."

CHAPTER FIFTEEN

HERMAN RHETT

A Fateful Choice
1948–1958

The sentry, who was black, looked at me in the cab
and said, "What the hell are you doing in there?" I said,
"Well, I'm reporting to base." He said, "Get your ass out of that cab right now."
So I did, and after I paid the driver, the sentry said, "You see that building
way down there?" I said yes. He said, "All I want to see going down
there is asshole and heels." I didn't know what he was talking
about, but what he really meant was run. So I ran.
That was my introduction to the Marine Corps.

The medal around Herman Rhett's neck represents his induction into the Montford Point
Marines Hall of Fame in 2001. It recognizes his
service as national treasurer, vice president, and
president of the Montford Point Marine Association, among other contributions. The patch on
his chest signifies that he is a life member of the
association, from the Washington, D.C., chapter.
(Photograph courtesy of Herman Rhett)

Like many of the former Marines in this book, Herman Rhett stayed connected with the service after his retirement. Over the phone, he told me how he had served four years as national president of the Montford Point Marine Association, from 1985 to 1989. His proudest creation, though, was a 132-page collection of thoughts, remembrances, and photos entitled The Montford Point Marines: Final Roll Call, *published in June 2004. He did it all himself. I ordered a copy from him, at a cost of forty dollars. "The book and the poster that goes with it were a labor of love," he said, "no grants, no corporate sponsors, just a willingness on my part to complete the project. I am extremely excited about the book and the hearts it has touched."*

"I was not a career Marine, although I spent ten years in the Corps and eleven in the Reserves, which is no place to be in any part of the service today. I went to school, studying computer technology, while I was in the reserves. I was among the last to be trained at Montford Point, in 1948. I served during the Korean conflict—although I never went there—and got off active duty in 1958. After Korea I was attached to the Marine Corps Air Control Squadron 21. I was the sergeant in charge of radio UHF and VHF equipment, all the mobile units."

Like many black recruits from the North in the 1940s, Rhett had some higher education. "I enlisted out of Boston. I went to Northeastern University for a year but didn't have the money to continue. So I went to the Army and took the signal corps test. They said the results would be back in about a week. Meanwhile, I went to the Navy recruiter, took an electronics test, and failed it by one point. However, they told me I could come in as a cook or a baker." Rhett was unimpressed. "I politely told them I wasn't going to join the Navy as a cook or a baker or a cabin steward to the officers. But as I walked out I saw this poster of a Marine in dress blues and I said, 'Boy, that is sharp.' So I walked into the recruiting office and the rest is history. I was nineteen.

"The recruiting sergeant was so sweet. He says, 'You come tomorrow morning and we'll swear you in with the other recruits and you'll be on your way.' I thought he was a really nice guy, white, but real nice. Tuesday morning I went down to the recruiting officer and there were twenty whites and one black. After he swore us in he turned around and went back in his office and when he returned it was like Dr. Jekyll and Mr. Hyde. He was Mr. Hyde then. His whole attitude changed immediately. We all looked at each other and said, 'What the hell did we do?'

"I went home and, lord and behold, there was a notice in the mailbox from the Army saying I had passed its test. They were going to send me to signal corps school and from there I would probably come out a second lieutenant, but instead here I was a Marine Corps recruit, not even a private. Anyway, those other recruits and I got on the train together, had a great time going down South, laughing and joking, and when we all got to Washington, D.C., all the white Marines went through the gate, and, when I started to go through to my car, the white conductor says, 'Where you going, Nigger?'

"I said I'm going to catch my train. I'm in the Marine Corps. And his comment, I'll never forget it, was 'I don't care who the hell you are. You're not going with them. You're going to stand here.' So I stood there by the gate waiting and next thing I know all the white folks on one Pullman car were getting off and, when they were all out, the porter, who was black, raised his hand, and the ticket agent said, 'See where that man is? You go there.' And I had that car all the way all to myself, from Washington, D.C., to Rocky Mount, North Carolina. When I got off there, I walked into a white waiting room by mistake, got ushered out of there, and they had me jump on the back of a Marine Corps mail truck that was going to Montford Point. This was October of '48. I was ready to go back to Boston then.

"Camp Lejeune was big. It had a lot of bases, Camp Geiger, New River, Hadnot, Montford Point, an air station. The truck dropped me at the base post office, where I was supposed to get another mail truck, to Montford Point, only a few miles up the road, but I saw a taxi cab and jumped in that. Well, that was the worst thing I could have done.

"See, I didn't know I was going to an all-segregated training base. I did not know that one bit. Everything started when I got to the gate. I laugh at it today but I didn't laugh at it then. The sentry, who was black, looked at me in the cab and said, 'What the hell are you doing in there?' I said, 'Well, I'm reporting to base.' He said, 'Get your ass out of that cab right now.' So I did, and after I paid the driver, the sentry said, 'You see that building way down there?' I said yes. He said, 'All I want to see going down there is asshole and heels.' I didn't know what he was talking about, but what he really meant was run. So I ran. That was my introduction to the Marine Corps.

"More than 565 black recruit platoons graduated from Montford Point. I was assigned to Platoon 18, the last one to go through, because the order to integrate the services came not long after. We had all black drill instructors. I remember one called Judo Jones. My D.I. was named Shaw. Gilbert Hashmark Johnson, who became famous, was sergeant major then, and when I got out of boot camp he called me in and said I was going to work in his office as a clerk, because I'd had that one year of college. But I told him, no, I didn't want to do that. I had joined the Marine Corps to fight; I didn't join to be a clerk.

"And he said, 'You have one or two choices. There's a building across the street. You can become an inmate there, or you can work in the office here.'" Rhett laughed. "I'll always remember that, as long as I live. Hashmark wasn't smiling when he said it, either. He just looked at me.

"In 1949 I was the first black to be sent to the personnel admininis-tration school at Parris Island. The jumping-off point was Yemassee, where all the white recruits came in, and it was all segregated at that time. I had one big stripe on my arm, I was a Pfc., and I wandered into the restaurant. The guy wanted to know what I was doing there. Mind you, I was in uniform. I told him I just wanted a sandwich or some-thing. And he said, 'You can't stand here. I can't serve it to you here, nor can you eat it here. You have to go around back, and I'll serve you there.' Just then two white Marine Corps sentries walked in and wanted to know what was going on. So I told them. One of the sentries looked at the restaurant guy and said, 'As of now, your restaurant is off limits to ALL Marines. If you're not going to serve this guy, you don't serve any of us.' I fell in love with the Marine Corps then.

"Another incident occurred after I got there when they sent us all to get haircuts because we were going to have an inspection. So we all go to the area set aside for the barbers and one of them looked at me and said, 'What are you doing here?' I told him I came for a haircut like the rest of the Marines. He said, 'We're not going to cut your hair. It's that simple.'

"Well, the next day the aide to the base commandant came and told me that in another week we were going to have Negro recruits on base and he wanted to inform me that those barbers had been terminated, and the general sent his apologies to me for what happened yesterday morning. And he gave me a three-day pass.

"When I graduated from school at Parris Island they sent me back to Montford Point and I helped deactivate it as a training base. Over 23,000 blacks had been trained there between 1942 and 1949. Those of us who stayed in got scattered all over the place.

"I did not want to go back underneath Hashmark Johnson so I managed to get myself sent to a school for electronic data processing at Henderson Hall, Headquarters Marine Corps, in Washington. But they would not allow me and two other black guys to sit in the same classroom with the white students, so the three of us were trained separately, and I think we had better training because everybody sympathized with us.

"When that period was over, all of us went back to Camp Lejeune to set up an automated data processing facility. After that was set up, I got transferred back to Headquarters Marine Corps, where I was enrolled in an advanced data processing school in Fort Lee, Virginia. I was the lowest-ranking Marine student in the class.

"I met my wife there. She was working for a colonel in supply. She was not in the Corps. We got married in 1951, just celebrated our fifty-third anniversary. We have three children. One son went in the Army, our daughter became an FBI agent, and our oldest son is a mechanical engineer in Amherst, Massachusetts."

After finishing the data processing school in Fort Lee, "I was sent back to Headquarters Marine Corps, and when I got back the equipment was already there and already functioning. We had eighty-column punch cards. One was called a status file and the other was a qualification file and they contained all your personal information.

"When the Korea conflict started and MacArthur called for the Marines, he wanted some that could speak Korean, because he needed interpreters. My first assignment back at Henderson was to go through almost a quarter of a million cards to find all the Marines who could speak and/or write Korean. And there were trays and trays and trays of cards, and of course I just got out of school, so I knew how to wire the machines to look for that particular code, and only ten out of a quarter of a million fell out that could speak or write Korean. I gave them to the sergeant major in charge of the facility. I don't know whatever became of those Marines, but I can well imagine. Wherever they were in the world, they were shipped out."

Herman Rhett went on to to work for NASA in its advanced computer research division in Cambridge, Massachusetts, taught computer-related subjects, and eventually helped monitor government contracts having to do with information systems. He received a Ph.D. in computer science from Boston University in 1980.

"Like a lot of the guys, I felt like I was fighting a war on two fronts in those years, but I found the best way to cope with discrimination was to ignore it, just ignore it.

"When someone asks about my military career, I kind of step in the background, because I came in at the brink of integration, and I did not encounter nothing like what these fellas who went before me did. Oh, yes, when I joined the Marine Corps I got pushed around in Washington, D.C., and at Yemassee, and all those little things, but that was nothing compared to what some of those Montford Pointers went through."

CHAPTER SIXTEEN

ELLIS CUNNINGHAM

First Sergeant
Iwo Jima, Inchon, the Tet Offensive
1943–1970

Everything was under fire. It was terrifying,
but war was for young people. Young men, they don't
give it much thought. Somebody's gonna get killed,
but it'll be somebody else, not me.

Ellis Cunningham and his wife, Lucille, got
married at her home in Charleston, South
Carolina, on December 27, 1953. Ellis went
right back to work after their wedding, but
they went on a honeymoon the following
June, to Washington, Philadelphia, and New
York. He joined up in 1943, and retired in
1970. This photo was taken at their church in
the summer of 2004. *(Photograph courtesy of
Ellis Cunningham)*

July 26, 1948: President Truman promulgates Executive Order 9981 banning color bias in the armed services. "It is hereby declared to be the policy of the President that there shall be equality of treatment and opportunity for all persons in the armed services without regard to race, color, religion, or national origin." The order also establishes the President's Committee on Equality of Treatment and Opportunity in the Armed Services.

Ellis Cunningham met his wife-to-be, Lucille, while he was recovering from a bullet wound at the naval hospital in Charleston, 110 miles from Parris Island. She was working at the Navy Yard. They were fixed up, he said, and it was "somewhat" like love at first sight. Lucille told me on the phone that someone asked to bring Ellis by to meet her and she said okay. "I didn't pay much attention, and then he brought this tall Marine in. I'll always remember that night he came to my house. He was wearing, what do you call it, this battle jacket, and it looked real good. He was limping. He looked real handsome." They got married three years later, more than fifty years ago. She has always called him "Cunningham." In 1995, he took her to Iwo Jima. "We stayed in Guam, at the Hilton," she said.

"I was shot in the upper part of my right leg September 28, 1950, just outside the wall of Seoul. I was with Item Company, 7th Marine Regiment. I was the sergeant handling communications for the company commander. It was early in the morning and we were moving to a point just outside the city. We'd been harassed all night with fire. We came under more as we were moving from there, and that's when I got hit, sniper fire, probably.

"The Korean War started in June, and by September they were

threatening our hold on Pusan, down south. The outfit I was in went around the other side of the island and landed at Inchon there to relieve some of the pressure on the people at Pusan. The First and Fifth Marines went in on September 15. It was a very difficult landing.

"My outfit came ashore five days later, in little boats with front ramps that came down. Our landing was not contested. We were relieving the Fifth Marines. They had taken a pretty good beating. The North Koreans didn't appreciate them coming; they were resisting a whole lot. It was only thirty-five miles to Seoul. The Fifth was not there yet but they were on the way.

"We took Seoul September 29, the day after I was hit, and went on north. I didn't actually see where the bullet came from that hit me. I heard it but I didn't see it. I was ten yards from my company commander. I spent the rest of the day getting back to the aid station. They brought us back to this little runway at Kimpo Airport. They had a planeload of wounded, most of us strapped down on stretchers. It was the first time I'd been on an airplane. We flew from Kimpo directly to Japan, Sasebo, and I stayed there until I was able to travel. Then they flew me to California. I ended up in the naval hospital in Charleston, South Carolina, until October 1951. The nerve system was shattered from my upper thigh on down to my toes. I couldn't move my toes. The first day in the hospital they put a cast on my leg in order to stabilize it, the first of three they tried. My foot had dropped down so I couldn't control it whatsoever." This did not deter Cunningham from wanting to fight. "When I got well enough to go back to duty, I went before an evaluation board and told them I didn't want to get out. They said the choice was mine." Cunningham chose to stay in, and skip more elaborate treatment for his injured leg. "It cost me a little problem moving around but it's no worse now than it was then, more than fifty years ago. I didn't want to be operated on. I met Lucille while I was at the hospital. We got married December 27, 1953."

Cunningham did not set out to join the Marines: "I started out in the Navy in 1943, and I was on my way from Camp Jackson, South Carolina, to Illinois to become a cook or a steward or a baker. I was in a group of forty-three blacks walking down the road on our way to the train when a Marine recruiter came along and told the petty officer in charge he needed thirty-two of us. He said, 'Okay, take 'em, you, you, you, and you,' just like that. I was one of the 'you's. That's how I got to Montford Point.

"I was born October 26, 1925. My father died when I was young, and I grew up on a hundred-acre farm with my grandparents. We farmed with mules, grew cotton and tobacco. I was five feet seven and weighed 116 pounds when I started boot camp. Other guys lost weight. I gained weight. I didn't find it difficult. I was used to working.

"After boot camp and an eight-day leave they put me in an ammunition company, which started forming the same day I got back. After additional training, we took a train to Treasure Island, near San Francisco, and from there we got on a boat and went to Hawaii, arriving in August 1944. We worked from the time we got there, receiving ammunition, separating it, repacking, then loading it back on ship—bombs, artillery shells, everything. We worked from the time we got there till January of '45.

"We got back on ships and left there on the nineteenth of January. Very few of us had any idea where we were going. Wherever the ship goes, that's where you go.

"Little did I know we were on our way to Iwo Jima."

Cunningham remembered being impressed by the size of the American force. "They had ships. As far as you could see, there were ships. We were in a large convoy and the ships we were with carried supplies. The primary landing force people were in transports, specially built for that purpose. We were on LSTs that didn't travel as fast. We left several days earlier than the transports, but they caught and passed us the day before we got to Iwo Jima.

"They were the first ones to land, on February the 19th. The first wave landed about nine o'clock. The first seven waves went in amtracs, then the people on my ship started going ashore. My ship went right up to the beach and I walked off right onto the beach. It had a ramp that opened in front. Everything was under fire. It was terrifying, but war was for young people. Young men, they don't give it much thought. Somebody's gonna get killed, but it'll be somebody else, not me. People dying right near you, and you hate to see somebody get shot, but you think that's not going to happen to me. A lot of people kept that attitude, although some people actually cracked up under the pressure. That's the way it went."

Once on the beach, Cunningham said, it was all action. "The beachmaster assigned everybody. We had ammunition loaded on pallets and strapped down with metal bands. We had to put steel matting down for vehicles to run on. Most everything was moved by bulldozers. They would pull the matting to where you needed a vehicle to go. We were under fire all the time.

"Eventually you got a foothold where you needed to be. We had to put big balloon tires on the trucks so they could operate. We didn't have to cut through the bank. I believe we landed on Beach Red. We weren't next to Suribachi but right close to it. It was never out of my sight.

"My company was divided into three platoons, each one serving each of the three divisions, the Third, Fourth, and Fifth. The Third stayed in reserve and the other two landed abreast of each other. We had ammunition for everybody—rifle rounds, flame throwers, grenades, machineguns, Browning automatic rifles, everything. Usually the troops came and got it themselves, but at times we'd bring it to them. If they didn't have a way to get ammunition up, one of our trucks would take it up as close as it could go, and we'd carry it the rest of the way.

"At night you'd dig in wherever you were. You maintained unit integrity as a squad; but you didn't want one shell to get all of you, so you'd spread out. We had M-1s. We lost a couple ammunition dumps that were hit by shells, and a couple fuel dumps that way too.

"Did we feel resentment because we weren't fighting with the rest? At the moment you really didn't give it much thought, because you'd look at what happened to that white guy there: He wasn't on Easy Street. He was getting his butt shot off. The hour I came ashore I guess there was 400 or 500 dead guys on the beach right at the edge the water. You didn't have any time for any racial animosity at that moment.

"We were on the beach the whole time. We had several guys shot, a few got killed, fewer than you would think. I wasn't hurt. Each unit put out its own security at night. One of our platoons was providing security at the runway at the first airfield we captured. At some point you will read about the last banzai attack on Iwo Jima. It was at this airfield. It was a black unit that they attacked. The guys fought very gallantly. It was early in the morning when the attack came. That's the only place where a few of our guys were awarded medals, for their performance that morning. Part of my unit was there; I was at the other end of the field. We were spread out pretty thin but the guys held their own. The Japanese learned where these air people were camping out and decided to attack and wipe out the pilots, but they had to come through the Marine unit to do it, and the Marines wouldn't allow that

"One of our guys, Charles Burnett, who's actually from the Charleston area here, got a Bronze Star. I've not seen him since we left Iwo Jima. I'm told there were other medals. Had to be a few Purple Hearts.

"We were there the whole thirty-six days, then went back to Hilo to get ready for the invasion of the mainland, and we were headed there when the war ended. We kept going and landed at the naval base at Sasebo; then, on New Year's Day 1946, we left for Guam, where we were put to work locating and destroying stockpiles of ammunition, collecting all the ammo that was left over from the invasion, scattered all over the island. It was very dangerous. We'd collect and load it onto trucks, then at night take it down to the docks and unload it onto a huge barge. It would take two or three days to do that. I don't know why they did it at night.

"Then they'd haul the barge way out to sea and push the ammunition off. It was artillery shells, mostly. I had reenlisted on Guam, and got promoted to corporal. We returned to San Francisco Mother's Day of 1946. There was no expiration date on our service contracts, which were 'for the duration and six months.' We went on back to Camp Lejeune and Montford Point. I'm not sure why I chose at the time to stay in, and haven't given it a lot of thought since.

"After the war they formed another anti-aircraft unit. We had two anti-aircraft units during the war, the 51st and the 52nd, and afterward they formed the 3rd and I was in that for a while, then they sent some of us black guys out to an ammunition depot in Oklahoma and I was there for a a little over a year.

"President Truman's order to integrate in 1948 brought a few little skirmishes. The Commandant had said at the start that a black would not tell a white what to do, and no one bothered to test that. And that was when you first started realizing you weren't sure if you wanted to keep the stripe or lose it, because if you were a sergeant or above, you still could not command white guys. If you couldn't tell anybody else what to do, you were out of a job. If you were a private or a Pfc., it wasn't any problem.

"My rank led to problems when I went to Japan in preparation for Korea in 1950. I had helped on the drill field at Camp Pendleton when they brought in three reserve battalions for training. I made up the schedule and supervised the execution of the training. They were all to be folded into the Seventh Infantry Regiment. We had a few regulars there to help mold them. The overall proficiency was poor, and it took a while to shape them up. Once the training was finished, everybody was assigned a permanent job.

"Well, I didn't get an assignment. We got to Japan and I didn't have a job. I was shuffled from one unit to another with nothing permanent until I got to Item Company with Captain Singwald. He told me what the problem was—I knew what the problem was—segregation. The captain didn't beat around the bush. He said, 'Sergeant, I'm supposed to send you to another unit, although no unit wants you.' The reason they didn't want me was because I was a black sergeant. So the captain told me, he said, 'Well, I'm not going to shift you around to another unit. I'm going to keep you with me.' He put me in charge of communications. He gave me six men, blacks and whites, to man the radio and telephones. I was in charge of telling them which platoon they would work with and when and where to put the wires out. Every time the unit stopped, you had to run a telephone line between the platoon leader and the company commander. It was my job to see that everybody got connected. That what I was doing until I got hung up with that bullet early that morning outside the wall around Seoul.

"Fast-forward now to Camp Lejeune, 1951. We're supposed to be integrated now, and I'm at the rifle range, and I'm in the butts operating the targets. I had a hundred people down there operating the target. I'm telling them how and when to operate the target, and I heard Mr. Martin, the warrant officer at the range house [a warrant officer is in between a noncom and commissioned officer], call the master sergeant, McCoy, at the firing line, and he said, 'Hey, McCoy!' 'Yeah?' 'What are them two little black bastards doin'? Where are they going?' McCoy said, 'Oh, I sent them back to the range house to get some new ammunition.' Martin said, 'Oh, I thought they were slipping away.' He saw them off in the distance going across the field. 'No,' said McCoy, 'I sent them.'

"When the two of them finished, I sounded off on the public address system, Cease firing! I wasn't supposed to do that, but I did. At that moment I didn't care what happened. I pulled the target so nobody could shoot at it. People up on the firing line didn't know what happened. I crawled up over the butts there and ran up to the firing line.

"By this time McCoy knew there was something wrong, and Mr. Martin knew there was something wrong, because this was highly unusual. I ran on up to the range house and got right into Mr. Martin's face. I told him, 'Now, Mr. Martin what you just did, you're lucky I'm not a hothead like some of the other guys that you would run into, because instead of talking to you, I'd have a fist in your face. I understand you're a Marine officer, but your conduct is far less than what a Marine officer should be. You know what the rules say about integration and what your conduct should be.

" 'If those two youngsters there were to address you the way you addressed them there could really be trouble.' I kind of dressed him down. He didn't like it at all. He said, 'Let's go see the colonel. The colonel would agree with me.' I said, 'All right, let's go.' He figured I was going to back down, but instead he backed down. He went on to say what he could to appease me. I said, 'All I got to say to you is, you better conduct yourself as an officer is supposed to and treat these guys the way they're supposed to be in an integrated unit and don't let me have any more problems out of you, because if you do, we will go see the colonel, whether you want to or not.' I had no more trouble out of him and very little trouble out of the other guys from that time on.

"In '65, they sent me to Parris Island to administrative school. When I finished they promoted me to first sergeant, so I stayed in until 1970, long enough to qualify for the retirement income that came with the rank. I got orders to Vietnam in 1967. I was there for the Tet Offensive, the granddaddy of them all, plus a few skirmishes. I was with the Third Division as first sergeant. I ended up at Dong Ha. Christmastime I was up at the DMZ, where a lot of fighting was going on. They were trying to run us out of there. I was back in Dong Ha for Tet. The barrage started about eight o'clock that morning and continued every fifteen minutes for the rest of the day until about four. It was a very scary time. About eight Army helicopters had landed a hundred yards from me the evening before, and they were all destroyed. The North Vietnamese never tried to come into our positions. They just kept shelling us."

Would he do it all again?

"I would recruit for the Marine Corps right now, to any youngster who asked me, so long as they're serious. The initial training is so strenuous they have to be real serious if they're going to choose the Marine Corps. The training, the strenuousness of it, is what sets it apart. It makes an impression that doesn't wear off easily. As for integration, I'd say that it took a while but, overall, the Marine Corps finally came around. I stayed in until 1970, then I drove a bus for the City of Charleston until I became a supervisor. For the last eight years I've been a volunteer at the naval hospital. I go around and talk to the patients, counsel them, and so on. I retired in 1970. I like to play golf. I'm not very good at it, but I love it."

Cunningham said his time at Montford Point left a lasting impression. "I still remember my D.I. His name was Shelburne. Hashmark Johnson? I really loved the man. Some people didn't but I did. I've heard him called a big ego, but you got to remember the time where we were and how conditions were—how Hashmark got to where he was. When you consider that, you understand he was a dedicated individual who pushed everyone to do his best. He wouldn't accept anything less than the best, if he knew you could do it, and he felt most people could, if they wanted to. It was his job to motivate you, and he went out of his way to do that. That's why I held him in high esteem. I will say I wasn't that close to him. He was the sergeant major by the time I got to Montford Point. There were other guys there too who were great leaders.

"I think the performance of the blacks who went through Montford Point was outstanding. You want to remember they were the first blacks ever admitted into the Marine Corps, and they did themselves proud. There were inequities. A white guy got paid twelve dollars a day and the black guy ten dollars, and a lot of blacks would not accept that. Give me twelve or don't give me anything. We're fighting the same war. Some of them excelled despite this and others destroyed themselves trying to correct it. So you had to be careful not to destroy yourself but fight to correct the wrongs. Some succeeded but too many destroyed themselves. You can't let hate destroy you."

CHAPTER SEVENTEEN

DAVID DINKINS

1945–1946
It Was Not Easy to Be a Marine

I didn't know from beans about the Marine Corps when
I went in. I didn't know there had never been any blacks in the
Corps before 1942. I didn't know about Montford Point or Parris Island
or anything, except I had a buddy who had a brother who was
in the Marines, and he just looked so spiffy in his uniform. He
was like everybody's idea of what a Marine should look like.
When I put the uniform on, I looked like a Boy Scout.

David Dinkins, who was all of five feet seven
and 128 pounds, struggled mightily to gain
admission to the Marine Corps via Montford
Point in the summer of 1945, and then the war
ended while he was still in boot camp. He left
the Marines in the summer of 1946 and went
to Howard University and then Brooklyn Law
School. He is shown here in 1989 giving a
campaign speech prior to winning election as
mayor of New York City. (*Photograph © James
Marshall/CORBIS*)

David Dinkins was mayor of New York City from 1989 to 1993. He was born in Trenton, New Jersey, July 10, 1927, and in August of 2005, at the age of seventy-eight, he was a professor in the Practice of Public Affairs at Columbia University in New York City and a Senior Fellow of the Columbia Center for Urban Research and Policy. A list of his involvements, from national chairman of the Black Leadership Commission on AIDS to membership in the Council on Foreign Relations, would fill two pages. I exchanged many phone calls with his cooperative and helpful assistant at Columbia, Carol Banks, until he and I at last made contact for this interview, which we conducted by telephone.

"The war ended while I was in boot camp.

"I had graduated from high school in Trenton, New Jersey, in May or June of 1945. I turned eighteen on the tenth of July and I went down and signed up. I figured the best way to stay alive was to be better trained and the way to be better trained was to be a Marine. The only way to get in was to enlist, and I had tried to enlist before I turned eighteen. But there were no recruiting offices in Trenton. So I went in succession to Newark, to Jersey City, to New York City, to Camden, and to Philadelphia. In each place I was told either we have our quota of Negro Marines or you have to go to the state of your residence.

"I pestered the guy in Philadelphia so much that finally he let me fill out papers and take the physical. Then they told me I had high blood pressure, which of course I didn't believe. I assumed it was just another obstruction. So I stopped at a doctor's office in Philadelphia. My blood pressure was normal. I came home and went to a family physician: blood pressure normal. So, I go back to Philadelphia. They take the pressure on my left arm, on my right arm, lying down, standing up, all ways. I was so persistent that I impressed the guy, and I was little—I was five feet seven and weighed 128 soaking wet. He wrote a letter to my draft board and gave me a copy, said this man passed the physical and requests the Marine Corps, put him in.

"So when I turned eighteen on July 10 I went down and requested immediate induction, and they took me on July 28. You don't come back home, you keep on going. There was one other black and maybe twenty-five whites, and we left out of Jersey City on the train. I wrote a ton of postcards to everybody I knew. I was so proud that I was going to be in the Marine Corps. It was fortunate I did this because I sort of memorized the route, writing it out so many times. We got to Washington, got off the train, and, when we reboarded, the other black guy missed the train, and he had my orders, all my papers. It was just the two of us who were going to Montford Point Camp. Everybody else was going to Parris Island.

"Every time the conductor came through collecting tickets I would say I'm with them. That worked until I had to get off to get a bus, I don't know where that was, Virginia, maybe, but that was my first real experience with Jim Crow. I walked in, they said you got to go around in back. Eventually I got to Jacksonville, this little town outside Montford Point, and I got the bus to camp and a great big black sergeant says, 'Where are your orders?' I launch into this long explanation of how come I don't have any orders, and he hit me. And I bounced. That was my introduction into the United States Marines.

"Boot camp was tough. We had black drill instructors, for the most part. We used to do the manual of arms with locker boxes. If you were being punished, they'd have you run around some clothes on a line. They'd say, 'Run around these clothes until they dry.' It was the first time I ever saw grown men cry. But, because I was little, I thought I could do anything. Psychologically some little guys tend to think they're tough. Anyway, I got through okay. I do remember one day the drill instructor came in and said, 'Get down on your knees and thank God the war's over.' Then he said, 'Now get up, nothing's changed.'

"When I finished boot camp, we went into malaria control. We'd go out in the boondocks and spray DDT all day. I had worked in a garage during high school so I filled out some forms, made it sound almost like I was a mechanic, and I hounded the CO and got transferred to motor transport. That was pretty good duty except they'd send out convoys any time and it always seemed like four or five o'clock in the morning, and I thought I could do better than this. Then I saw these guys in pressed uniforms. I said, 'Who the hell are they?' It was H&S [Headquarters and Service] company. So I finally wrangled my way into H&S and they made me assistant file clerk.

"Well, the file clerk was a corporal, and when he got discharged, I got to be really important, because I was the only one who knew where the records were. The war was over, but I was going to reenlist because they were sending people to radar school. They would give you a bunch of money and some liberty and whatnot. For a while there I was driving for the executive officer of the base, and he said to me one day, 'You know, if you want to go to college, you shouldn't reenlist.' I had no notion of going to college at that time, but I didn't reenlist.

"So I got out in August of 1946 after thirteen months in the Marine Corps. I didn't want to go to college, but my parents sort of insisted. I said, it's too late, I can't get in, this is August. But my stepmother knew the fellow who was in charge of veterans at Howard University—he had been a classmate of hers in the Class of '29—and he said, No problem. So I got into Howard in the fall of '46 and, as they say, the rest is history. I majored in mathematics and graduated in 1950 magna cum laude. I won a math fellowship to Rutgers, and went there for a semester until I stayed up half one night working on a problem. Next morning I went to class and went to the smartest person, and I said, 'What'd you get for Number 2?' She said, 'Pi over 3.' I said, 'Gee, that's what I got.' Then, as sort of an afterthought, I asked, 'How long did it take you?' She said, 'About 20 minutes.' Then I knew math was not for me. So I decided to go to law school. You don't need to be too smart to be a lawyer.

"I got married in August of '53 and started law school in September. I was a Dodger fan, so I went to Brooklyn Law School. I had met my wife, Joyce, at Howard. She finished in '53; I had finished in '50. I had to wait for her to graduate so we could get married. We have a little boy, David Jr., who's fifty now, and a little girl, Donna. She's forty-seven.

"I didn't know from beans about the Marine Corps when I went in.

I didn't know there had never been any blacks in the Corps before 1942. I didn't know about Montford Point or Parris Island or anything, except I had a buddy who had a brother who was in the Marines, and he just looked so spiffy in his uniform. He was like everybody's idea of what a Marine should look like. When I put the uniform on, I looked like a Boy Scout.

"I do remember they had no black officers and I remember hearing how some guy got commissioned as a lieutenant or a warrant officer and the story went that he was commissioned one day, marched in a parade the next, and was discharged the following day. And today you got blacks who are generals. We've come a long ways. The country has come a long way. I guess the most graphic example is Colin Powell, who was, you know, in ROTC [Reserve Officers Training Corps] at City College of New York and went on to become head of the Joint Chiefs and then Secretary of State.

"It's pretty clear we've come a long way. I remember when I was in college at Howard in Washington, D.C., in the 1940s and some of my classmates had shrapnel in their bodies but couldn't go to the movies downtown or shop on F Street if they were going to try clothes on. This was so illogical, given that we had fought this war to end all wars. And yet, in many instances in the South, Italian and German prisoners of war were treated better than were the black soldiers who were guarding them. So we've come a long way. But we've got a distance yet to travel."

PART FIVE

WOMEN MARINES

The Fight for Equality

WOMEN HAVE SERVED unofficially in the military through-out American history. They gained an official role in the U.S. military with the formation of the Army Nurse Corps in 1901 and Navy Nurse Corps in 1908. Over 300 women served in the Marine Corps during World War I, "freeing a man to fight," as the slogan went. Altogether, more than 35,000 women served, most of them nurses. By World War II, there were over 400,000 women in service. About 2,500 women served in the Marine Corps during the war. The first recruit training for women was held at Hunter College in the Bronx, New York, and it moved from there to Camp

Hadnot, part of what was to be Camp Lejeune. Today, roughly 350,000 women serve in all branches of the military, nearly 15 percent of active duty personnel.

During Operations Desert Shield and Desert Storm, in 1991, more than 33,000 servicewomen were sent to Southwest Asia, according to the Women for Military Service in America Memorial in Arlington Cemetery. Thirteen servicewomen were killed and two were taken prisoner.

Women are not supposed to serve in combat, but in settings such as Iraq and Afghanistan, where there are no front lines, they frequently encounter threats and danger. As of February 2006, more than 260 had been wounded, and 36 had been killed, mostly by IEDs (Improvised Explosive Devices).

Women began training at Parris Island in 1949. Today they make up the Fourth Recruit Training Battalion, totally segregated from the three male recruit battalions. Over the decades the training of women has evolved from relatively easy physical demands and classes in how to apply makeup, to an arduous training cycle very much like the routine followed by the men. In addition, as the following chapters show, women had to challenge and overcome the rule that, if they got pregnant, they had to leave the Corps. Today, about 3,000 women recruits go through Parris Island in a year. The dropout rate is 12 percent. There are 9,500 female Marines in the Corps out of a total population of 159,000. While men train at both Recruit Depots at Parris Island and San Diego, women train exclusively at Parris Island.

Not all the women in this segment served on the drill field, but all their stories are fundamental to the development and evolution of training in the Marine Corps.

CHAPTER EIGHTEEN

DENISE KREUSER

Sergeant Major, Parris Island
1978–2005

I also used to tell them,
"Don't go out there and be bimbos.
Be Marines. Be a professional."

Denise Kreuser was sergeant major of the Fourth Recruit Training Battalion, the only unit where women undergo recruit training in the Marine Corps, at the Depot in Parris Island. A career Marine, she joined in 1978 and saw a great many changes for women in the Corps before she retired March 2005. "Being a D.I. was the best thing I ever did as a Marine," she said. *(Photograph courtesy of Denise Kreuser)*

*I spoke with **Denise Kreuser** in the office of her boss, Lt. Col. Kim Johnson, at the Women's Recruit Battalion on Parris Island in April 2004. She was straightforward and candid, looking forward to retirement the following year. Her Marine career spanned great changes for women in the Corps. She and her husband, Clint (Chapter 24), also a sergeant major, were both serving at Parris Island. They lived off base, in nearby Beaufort.*

"Women first went to the firing range at Parris Island in 1986, although they had trained here since 1949, the same year blacks were integrated into white units. Women started off in the Marine Corps in the early 1900s and World War I with administrative jobs, 'freeing a man to fight,' I guess you could say. I think that lasted only two years and then they were disbanded. During World War II, they recalled women and we came." Kreuser said the training was much different for women at that time. "They took women through how to wear the uniform, and some drill but nothing like what they receive today. I think there were, like, 20,000 in the Marines at that time—female reserves let in to handle administrative jobs again, but I think they did more—working on planes and doing all kinds of things to free men to fight. The Marine Corps did not have nursing. The Army and the Navy had that.

"I was born in Jersey City, and we moved around. I graduated from Sussex County Vocational Technical High School in 1978. My brother had been in the Marine Corps seven years and I wanted to join, but he didn't want me to come in. He told me I should go into the Navy but, every time I went, the Navy recruiter was not there. And the Marine recruiter was always standing at the top of the stairs, looking down, saying, 'Why don't you come see me?' and I said, 'I wanted to go in the Marine Corps, but my brother said, No.' Finally, later, I went back once more to see the Navy recruiter and he still wasn't there. But the Marine recruiter was, and we sat down for about three hours, and I went home and told my mom I was going into the Marine Corps. This was about three months after high school. My father had died and I had six brothers and two sisters and my mom was raising us herself. There wasn't any money to go to college. I enlisted in September 1978.

"I laugh all the time with my drill instructors today about what it was like then. It's so great for them now, but, when I joined, training was not what I expected. I had grown up with brothers, and I figured it was going to be challenging. But when I got here, it was a little different: We were taught how to put our makeup on. There were certain colors that we had to have on our eye makeup, and the lipstick had to be a certain color. They taught us how to iron our uniform, and we went through a lot of inspections on our makeup and our uniform appearance. We never got to fire on the rifle range. Everywhere we went we walked with our oxfords and we had these blue slacks and a powder blue shirt and blue sweater. We had a blue cover [hat], call it a garrison cover, nothing like our male counterparts.

"There was some physical fitness but nothing like today. We did some PT, ran about a mile and a half in those plain white Keds, flat sneakers. I had shin splints the whole time I was in recruit training, I mean bad shin splints. We didn't do the Obstacle Course or the Confidence Course like they do today, didn't do any of that. We didn't get to fire the weapon. We got to pass it around and go, 'Oooh, ahh, an M-16 A1.' We went to the gas chamber to put on gas masks, too, and that's all I can recall. There wasn't a whole lot of field training.

"Not much had changed at that time since 1949. It was all about being a female. We actually had a tea party, I remember, right before we graduated. My drill instructors today think that's hilarious. We had a tea party, had the white gloves, the whole nine yards. It was like making hostesses. We've come a long way. We had female drill instructors, but the person who taught us our drill and most of our classroom instruction was male. I know I never wore the blue stuff and the red lipstick again after recruit training.

"Out in the Fleet when I first joined, females were going into what was called the Utilities: generators, water, motor transport, and supply. When I checked into my first duty station, my gunnery sergeant told me he had never had females in his shop before. So I was the first

in that unit. When I checked in and stood in front of my gunnery sergeant, I was scared just like any other young Marine would be. And he
told me—and this was the greatest thing he could have done for me—
he told me he didn't want any females in his shop but as long as I was
going to be there I was going to drill with the rifle, I was going to shoot
with the rifle, and I was going to do everything the males did in his
shop. And then he gave me, 'Do you understand?' And I was like, 'Yes,
Gunnery Sergeant!' Because that was exactly what I wanted. Some people might say, well, he harassed you. No, he didn't. A. C. Karp. I remember his name. He was just my mentor, the best thing that ever could
have happened for me. So, anyway, I qualified on the M-16 with the
gunny. I shot Expert almost every time. He was pretty gruff with me,
which was just what I wanted.

"That was the Second Medical Battalion at Camp Lejeune. His leadership was awesome, unique for me because he treated me like all the
other Marines. Consequently all the males treated me equally, they
became like brothers, so I never encountered any problems. There was
probably harassment in some sections, like everywhere else in our work
force, but I didn't face any because I was lucky enough to have someone of his caliber. There were some people that treated you the old way,
but a lot depended on how you performed. If you acted the damsel in
distress and said, I can't, I can't, then they treated you accordingly, but
having all those brothers I guess helped me adapt a little better when
I got my first duty station.

"Karp retired several years ago. I ran into him here later. He came
and visited me. He said, 'Oh, you shithead, I remember you.' He actually called me that, and here I was a sergeant major and an instructor
at the D.I. school. But he had the biggest impact on my Marine Corps
career; he was the best thing that could have happened to me.

"I came down here as a drill instructor from 1985 to 1987. The change
came in 1986. This battalion used to be called the Women's Recruit
Training Command. We were not attached to the regiment like we are
now. We were separate. In 1986 we fell under the Recruit Training Regiment. There are now four battalions on Parris Island. There were three
men's and they made us the Fourth Recruit Training Battalion. Female
recruits were then required to fire the weapon and females were

required to carry the NCO sword, whereas when I left recruit training and went out into the Fleet, female Marines did not qualify on the rifle range nor were they allowed to drill with the sword.

"So I was ten years ahead of the curve with the rifle. I didn't realize that till I came down here as a drill instructor, where the new concept was, Everyone is a Marine, every Marine is a rifleman. Women are still not allowed in the infantry or artillery, and today in Iraq, women are in motor transport, supply, utilities, support units, but you never know where the front lines are going to be.

"On my first tour, we wore red cords to symbolize we were drill instructors. The Smokey campaign cover didn't happen until the '90s. Female D.I.s were still wearing soft covers up to the '90s. I once had a corporal in my shop. She was with the engineers but she wanted to be a drill instructor, and she asked me one day, she said, 'How come you all don't wear the Smokeys?' Well, the female drill instructors had taken a vote and they didn't care whether they wore them or not. But this corporal thought it was important, so she wrote the Commandant of the Marine Corps, General Krulak, and said, 'How come we can't wear Smokeys?' And he said, 'Well, they took a vote they didn't want it.' And she wrote back and said, 'Well, we didn't get to take a vote when we increased our physical fitness test from a mile and a half to three miles.' And the Commandant wrote back and said, 'You're absolutely right.' And he changed it. I mean, that was a corporal writing to the Commandant. And that's how we got the D.I. covers. I was a D.I. at the time, but I didn't recognize the importance of it. But then when I came back in 1990 as a sergeant major, I realized that this was something that should have have happened to us many years ago.

"When you're young, you don't notice those changes because you're focused on growing up. I was focused on my own little life. But as I became a staff sergeant and a gunnery sergeant, I began to notice how much better it was for our female Marines. We'd started making changes in '86, but when I came back in '90 it was equal across the board. I did two tours on the drill field, then got assigned here in 2001. By then the women did every bit of training the men did, Obstacle Course, Confidence Course, pugil sticks, the bayonet course, and martial arts instruction. There's a few things, such as no pull-ups for women, because of the physical differences, but I got some female drill instructors today who can do more pull-ups than the men.

"So women get the same training as their male counterparts do and they're fired up about it. The first female recruit to fire a weapon here, Private Lobo, shot a 246 when 250 was a perfect score. They've got her poster hanging up in all the squad bays. There's a lot of males out there who will say there's females who can outshoot them, there's females who can outrun them—it's good, it's equal.

"We teach our female recruits to kill, even though they may not be 03 [Infantry]. Every Marine's trained to be a basic rifleman, so we teach the females the same things men are being taught. We teach them, Hey, sometime you might be next to a male counterpart. You might have to save his life. We teach that to females. We're there to kill and protect our buddy, and if you can't fight to save your buddy, who's going to do it?

"We had some wives go in the squad bay the other day and they were horrified to see three drill instructors all yelling at a recruit at the same time. They thought that was bad. After all, they're females and didn't we think that was kind of harsh? And our answer was, No, that was good. Because that was gonna toughen her up. She's going to go out in the Fleet and run into some male Marine who might yell at her, and then what? She's gonna break down in tears because somebody just yelled at her?

"So we put the pressure on them kind of like when we had pressure put on us at drill instructor school. Of course, we were Marines, and these are recruits. They're not Marines yet.

"This is why the Marines are unique in the American military: When recruits get here, we basically take everything about them away from them. We strip them of all individualism, starting at ground zero. From the get-go we talk about teamwork. They never do anything by themselves. If the person next to them messes up, then they both mess up. If they dislike another recruit, then we make them bunkmates. It's just all about teamwork and never letting your buddy down. If one fails, we all fail. I think it follows them when they go back into the world as well. Some people might say it's brainwashing but I think it's just what we instill in them, honor, courage, and commitment, the core values: Marines never lie, cheat, or steal, just what's right for the right reasons. Of course you know down the lines there's Marines who will stray, but I think the majority stay with their training. We keep reiterating that. And then there's discipline, obedience to orders. One of the things the Marine Corps prides itself on is putting a better citizen back out there.

"In the Fourth Battalion here at Parris Island, we have eighty-nine drill instructors. I'm in charge of all of them. And there's only fourteen or fifteen female sergeants major in the entire Marine Corps.

"Why do women join the Marines? Most of them say they came in for the challenge and a lot of them will say because they wanted to belong to a team. The Marine Corps prides itself on being a team, a family. We take care of our own—that's one of our little sayings. I think they want to belong to something that's special. They have a lot of different reasons for joining.

"Some, or a lot of them, do seem to be escaping from something, as Colonel Malachowsky says. A lot of them come from bad backgrounds, sexual and other abuse, foster care, and they want to get out. The recruiters say they don't actively recruit women because the women walk in and volunteer. I think that's kind of funny, because we're toughest and have the least amount of females of all the services. I always thought it was odd they pick the hardest one. I've been the only female in my shop probably the whole time I've been in the Marine Corps.

"There is total segregation in recruit training. Men and women never come together till they go on to their schools. We do have male instructors teaching swimming, but for troop movement it's better to have females in charge so the recruits can have good role models. Like the colonel said, some females come from bad backgrounds, and I think they need that strong role model, a female professional to teach them. The other thing with the female recruits too was a safe environment, having them in a safe place where they don't have to worry about anything else, like men. There's been a lot of issues with our being separated from the men. We may be the only battalion in the U.S. that is totally segregated. Males and females train together in the Army, but I think we're doing it the right way here. Our female recruits get all googly-eyed with just the male instructors, never mind being next to an eighteen-year-old male recruit. Everybody's hormones are already going crazy in this place, so it's best to keep them separate.

"Following my tour with the Second Medical Battalion, I went to Okinawa. I didn't meet my husband, Clint, till I went to the drill field in '85. In Okinawa, I purified water, worked with generators. See, when you set up camp out in the field, you have different support units that give you power, water, and so on. My MOS was Utilities. I was there about a year, came back around 1981 and was at Cherry Point, North Carolina, air station until 1985. Unbeknownst to me my staff NCO and my lieutenant had put on my fitness report that I should be a drill instructor. I was scared to death. Drill instructor school is a scary place. The worst part is the first three weeks, when it's almost like being a recruit. You can't get the time management down. There's so much pressure. Your uniform's gotta be perfect, you got to be perfect plus, by the way, you gotta write this and you gonna teach that. There's just a massive amount of stuff being dumped on you and you don't think you can do it, just like a recruit thinking, 'I can't do this.'

"Sometimes I think you get more stressed out mentally than you do physically, even though the physical part was hard over there, oh, my gosh. As a drill instructor, you've got to be in 100 percent better shape than the recruits. You do a whole lot more exercises so you're stronger than they are. I can't tell you how many miles we ran, but it was a whole lot faster than a normal Marine would run. I couldn't even do today what I did then. But by the time you graduate you feel like you can do anything. School lasted twelve weeks, three months.

"I met Clint in D.I. school. We got married in 1986. He was in my squad. He told me later that as he sat behind me he said to himself, 'I'm gonna marry that girl.' I didn't give him the time of day, actually. We just became friends and it went on from there. When you're in D.I. school, you're broken into six or seven different squads, and you develop strong bonds with people because there's competition. Females and males train together, PT, do everything together, even though the recruits are segregated. So our squad studied as a unit and practiced individual movements with each other. You take turns explaining things over dinner. You have to teach to an instructor everything you're going to teach a recruit, so you teach it first to each other in your squad.

"I loved being a D.I. It was the best thing I ever did as a Marine. You have such an impact on the Marine Corps with the recruits. You get 'em from Day One and they're just lost. And you can see the changes as you go. It also depended what part of training they were in. I could always tell when we went to the rifle range—seeing their confidence level. You'd just go out to the squad bay one day and tell them to get on line and they would all just move as one, and I was like, 'Wow, they're there.' And then they just kept progressing till they graduated. It was an awesome feeling to see your work. It's almost like you built something. And you could see the end result when they graduated. It was a good feeling.

"You're transforming their lives. I always tell my drill instructors every one of them is going to impact that recruit's life one way or another, and every recruit's going to be different. Recruits will always remember their drill instructors. I still have officers writing me from my first tour in '85, and sometimes I get the tours mixed up. I'm like, I don't remember who this is. But they remember every little thing about you, everything you told them. You'll run into them and they'll say you said this to us. I used to always tell them, 'Hey, you pull your own load. Never have a male Marine pull your load. You figure out a way.' And I had somebody come back and repeat that to me. I also used to tell them, 'Don't go out there and be bimbos. Be Marines. Be a professional.' "

Here she laughs and recounts an anecdote about a meeting with Iron Mike Mervosh and a couple of her men during the recent drill instructors' reunion:

"I introduced two of my lance corporals to Iron Mike Mervosh during the D.I. reunion and he said, 'Oh, you work in the Fourth Battalion with all those broads.' And my husband, who was standing there too, he's like, 'Did you hear what Iron Mike just said?' And I just looked at my two guys and I said, 'Lance Corporal Melser, Lance Corporal Prewitt, you better not even think about calling us that name.' And they're like, 'Oh, no, Sergeant Major.' Some old-timers used to call us BAMs, broad-assed Marines. When I came in they were still calling us that, and somebody said one day that's not what it stands for; it stands for Beautiful American Marines. But I wasn't sure I liked that one either. They called us WMs for a while too. But I just wanted to be known as a Marine.

"The drill instructors work over 100 hours every single week. It's phenomenal how hard they work. I think about it because I'm older now and I thank God it's them and not me. Let the young ones do it, because I couldn't do it again. You do a three-year tour here, and one of those years is spent in what they call a quota. It's kind of a break and you don't have the stress and the hours you would have as a regular drill instructor. You teach field stuff, all support. Normally you try to get the instructors out when they've worked about three platoons, which is almost a year, but sometimes people will handle four and five platoons before they go on quota.

"There are three companies in the battalion, and each company has a first sergeant, which is one rank down from me. The first sergeants run the companies, and I'm responsible for all three companies. I always keep a finger on the pulse of the battalion. Each company consists of two series, and each series team consists of two platoons, all under a gunnery sergeant, who has two senior drill instructors and each of whom has two Green Belts under them, like a second and third Hat, or junior D.I.s. So you have three D.I.s assigned to a platoon, and the gunnery sergeant keeps them out of trouble, makes sure everything runs smoothly.

"Chris Henning, who took part in the cadence counting competition at Traditions during the D.I. reunion, is a series gunnery sergeant. She actually did a whole drill card, which takes a platoon through the marching sequence. You get graded on an initial drill with a platoon of recruits to see how far they have come, and then there's a final drill toward the end. The drill instructor will go out there with a whole bunch of movements on the card and count cadence as she takes the recruits through the movements. This is a big deal, and they compete for trophies. Cadence counting is one place in the military where there's some room for individuality. You can put your own little twist on it. Some of them sing. When you learn, they tell you to base your Left, Right, Left, Right, on the movement of your car's windshield wipers, 120 steps per minute. Most people probably learned it from the their drill instructors when they were recruits. Henning did the whole card in her head. She's a little motivated. She calls cadence all the time on the catwalks."

Kreuser gave a quick lesson on Marine Corps terminology. "Catwalks are sidewalks, the ground is the deck. A cover is a hat. A hatch is the door. The water fountain is the scuttlebutt. This is all naval terminology, actually. The drill instructors have a bunch of little lingo things all their own here at Parris Island. Sneakers are go-fasters. Your hands are your paws: Show me your paws. The globe thing that goes around your waist they call the glow worm. A pen is an 'ink stick.' A thermometer is known as a 'rectal rocket.'"

Asked to summarize her opinion about women's place in the Marine Corps, she said, "I think we're right where we should be. I think we're pretty much on a level with our male counterparts now. We can always improve, but, as far as the recruit training process goes, I think we're where we should be.

"What am I planning to do next? I'm planning to retire next year, in 2005, when I'll have twenty-six and a half years in. I was apprehensive about it at first, but I think I'm ready. You never want to leave the Corps. My husband creates displays of ribbons and medals that he sells over the Internet. I haven't helped him with that but I will when I get out. I'm forty-four years old, and I'm not going to become a cop, which is what a lot of retired Marines do. I'm too old, I think. I might get into security or something like that. We're moving to Jacksonville, Florida.

"Did we want to go to Iraq? Oh, yes. The first war, the Gulf war, when that happened my husband and I had just come here and we missed it. And then here it came again in Iraq, and we missed it again. You feel like you're missing a part of the Marine Corps. We have so many people went over there. I know there's a lot of female Marines over there, in motor transport and supply."

Note: Sergeant Major Clint Kreuser retired in July of 2004, and Denise took over his position with the Support Battalion until she herself retired the following March 19, 2005. Reached in mid-May of 2005, Denise was asked if it felt strange being out of the Marine Corps. "Yeah, it does," she said. "You feel like you don't belong to anything." They were thinking of staying in Beaufort.

CHAPTER NINETEEN

DORIS KLEBERGER

1943–1966
"I Had to Do Something"

[Women] were not pampered, but were
expected to measure up to the standards of the Corps
in every aspect other than combat readiness. We were not
expected to go to combat. We weren't trained for
that. Now the whole world has changed.

Doris Kleberger at Quantico, Virginia, during
the winter of 1964–1965. Too young to qualify
as an officer when she signed up in February
of 1943, she went to "boot camp" at Hunter
College in New York instead. That fall she
became an officer candidate at Camp Hadnot
at Camp Lejeune, North Carolina. She eventually served several months as CO of the
Women's Recruit Battalion at Parris Island.
She retired a lieutenant colonel in 1966. *(Photograph courtesy of Doris Kleberger)*

Doris Kleberger retired at Camp Lejeune, North Car-olina, on November 30, 1966, after twenty-three years in the Marine Corps. Since then, she told me over the phone in the spring of 2005, she has "enjoyed life." She began her Marine Corps career as a private and ended as a lieutenant colonel. She was a championship golfer, whose ambition was to play golf courses "all over the world" when she retired.

"I live in Santa Rosa, California. I am single and have never married. I was born in St. Louis, Missouri, on July 3, 1922. I attended grade school, high school, and college in St. Louis, graduating from Harris Teachers College on January 29, 1943, with a degree in education. I had become very interested in women's part in the war effort and their increasing entry into the work force.

"My brother had been in the Navy from 1936 and then the war started and there was more and more war news, so as I was finishing college, I just felt I had to do something. When the Army and Navy opened their ranks to women in the latter half of 1942, I was envious but still had about three months to complete college. I received my degree on January 29, 1943. And two weeks after I graduated, the Marine Corps announced on February 13, 1943, that it was going to finally, reluctantly, take women into its ranks. The recruiting slogan was 'Be a Marine, Free a Marine to Fight.'

"I wasn't old enough to be an officer. I was just twenty. On February 22, I was sworn in and attended the first recruit training class at Hunter College in the Bronx, New York. Upon graduation, I was assigned as a recruiter in Syracuse, New York. In late summer or early fall, by which time they had moved the recruit training and the officer training down to Hadnot Point at Camp Lejeune, they announced they were going to take enlisted women instead of just officer candidates straight from college or the business world or whatever. So I applied for that first class of enlisted to go to officer training and made that

and went down there. I think it was about September of '43. The first
seven classes had been composed of eligible civilian women.

"Officer training consisted of phys ed and marching, learning the
history of the Corps and administration, that sort of thing. As I recall,
we had only female instructors. What did we do as Marines? We freed
the men to fight. The attitude was we were taking over the jobs we
could do here in the States so men could be released to go to combat.

"I was commissioned as a second lieutenant on November 29, 1943,
and was transferred to the radar school at Corpus Christi, Texas. Upon
completion, I was transferred to the Marine Corps Air Facility at Santa
Barbara, California, and served as station radio/radar officer until the
end of the war. The degree of sexism you encountered varied with individual men. Most were good and accepted the fact that women were
in, but some were a little bitter that they had to go off and possibly lose
their lives, but that was an individual thing. There were a lot of men
who couldn't accept the fact that a woman could do the things they did.
A lot depended too on how the women deported themselves. I thought
that overall the camaraderie at the air station was wonderful.

"When the war ended there was no opportunity to stay in. I was
released from active duty and retained as a first lieutenant in the
Marine Corps Reserve. I went back to St. Louis, then out to Santa Barbara, then up to the University of California at Berkeley, where I took
advantage of the GI Bill. I had studied biology in college, but when I got
to Cal my senior professor recommended that I switch to parasitology.

"I had about two and a half years in toward my Ph.D. when Colonel
Katherine Towle, director of Women Marines, wrote asking if I would
be willing to come on active duty for five months to serve on the staff
of the Officer Training Class at Camp Lejeune. Women had been
authorized, by law, to serve in the 'regular' services, and so the Corps
reactivated recruit training at Parris Island and officer training at
Quantico. So I did that, and, when I finished the course, I found I
could not go back to civilian life. I requested and was granted authority to transfer to the regular Marine Corps, but I had to go back in as
second lieutenant.

"I went on to spend quite a few years as administrative assistant to
Colonel Towle and her successor. In 1956 I became commander of a

women Marines company in Hawaii until April of 1957, when I reported to Parris Island as a major, executive officer of the Women's Recruit Battalion, as it was called then. It was much smaller, much different from what they have now. After a year or so the commanding officer was transferred and another officer, a lieutenant colonel, was coming from the West Coast to be CO, but she had a heart attack and so I became the CO for a number of months.

"The emphasis was on training women as GOOD Marines but not in any way for combat, or as the men were trained. Our women staff NCOs were not 'D.I.s'—they were platoon sergeants. The women were trained to march, to appear in parade formations, and they excelled at it. They were not pampered, but were expected to measure up to the standards of the Corps in every aspect other than combat readiness. We were not expected to go to combat. We weren't trained for that. Now the whole world has changed.

"There was a large field where the new recruit training facility is now and I used to hit golf balls there after work. I'd love to go down some day to see what they've done to my practice field. I was the All-Marine women's golf champion in 1959 and '60. When I was stationed in Hawaii, golf was THE thing to do, and at Parris Island, the area where the women Marine buildings now stand was a nine-hole course. It was directly behind the Women Officer Quarters. I would hit balls in the evening, go out, pick them all up, and start all over again. I did poorly in the All-Marine tournament in 1958, at El Toro, but I won the following year, when it was at Cherry Point. The next summer I was stationed in D.C. in the 5th Reserve District, but took leave to play in the All-Marine which was at Parris Island, in 1960. I won that. There was no question—I knew the course, having just left it after almost four years. That was the last All-Marine for the women. Just not enough participants.

"I served until November 30, 1966, when I retired at Camp Lejeune as a lieutenant colonel. I requested no parade, only that I be retired at morning colors with the color guard being composed entirely of women Marines. The hardest part for me was saluting the flag while wearing a Marine Corps uniform for the last time.

"The story they printed when I retired said I was going to travel the world playing golf, but I have to admit about seven-eighths of the world still has not seen my clubs.

"I have belonged to a golf travel club and still try to fit in as many of their trips as possible. I mostly play in the U.S., on courses in California. I have played many, many Canadian courses, mostly in western Canada. I have played a great deal in Hawaii. My biggest thrill was playing St. Andrews in Scotland. On that trip, I also played Turnberry, Prestwick, Gleneagles, and Carnoustie. My golfing travel days are winding down, now that I am in my eighties.

"As with other aspects of life in this world, things have changed. Women today have a greater opportunity to expand their possibilities and have the opportunities laid wide open to them. I won't say I envy them, but I can say, for my generation, I don't think we would change the way we had it and I hope those of today are as thrilled with their Marine Corps careers as we were, and continue to be.

"I admire what women Marines do today, and I've met them at some of our conventions. I'm not sure about the training because I come from a different era. I liked it the way it was. We served as we were needed, and as our place was to be, and I was very proud of my career in serving but I never had any envy. Now, as I look back on it, I don't think, Gee, I wish we could have done this, I wish we could have trained for combat. No. I have no nostalgia for that sort of thinking at all. I don't know if that makes sense but that's the way I feel.

"So much has changed in the whole world, civilian as well as military, and I see no reason why women shouldn't be trained for combat. And obviously they enjoy that aspect or they wouldn't be going in. I have a young friend who is now a captain, I met her out here in the Santa Rosa area. I knew her mother and her grandmother. She is a captain and a pilot, and she's stationed at the air station at Beaufort and I'm very very proud of her, so it's a whole new world. But back in the days when I was a radar officer I don't recall ever having any thoughts about wishing I could be a pilot or do what the men get to do.

"I'm proud of my career and I think we did what was asked of us, served the way we could then serve. So I have no regrets."

CHAPTER TWENTY

MARY SUE LEAGUE

Lieutenant Colonel
1960–1970, 1970–1985

They had never had a request like that,
and they didn't really know what the hell to do
with it, to be honest with you.

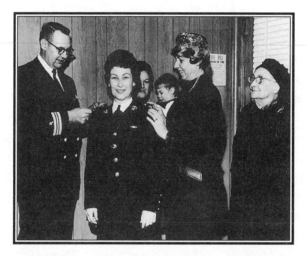

Mary Sue League is recommissioned January 1972, after regaining admission to the Corps. Her husband, Chaplain William League, USN, is on the left. Others in the photo are Maureen Walsh, a friend, holding Mary Sue's son, Steven, a year and a half old and the cause of it all; Nannette Beavers, a former CO, and Ms. Beavers's mother, Leola Hopkins Beavers, a World War I veteran. *(Photograph courtesy of Mary Sue League)*

After speaking with Mary Sue League on the phone, I visited her and her husband, Bill, at their home in Fairfax, Virginia, and we reviewed the things she had told me. I took a couple of photos. As she said, she did not normally talk about her experiences in challenging the Marine Corps regulations. "It was a very bad time in my life," she said, "having to fight an institution that I loved so much in order to get them to change their stupid dumb regulations so that women could be on active duty and have children. There was no reason for it, and I still get upset when I talk about it." She stayed calm, however, and she and her husband and I had an enjoyable visit.

"I was the first woman in the history of the Marine Corps to challenge the regulation. It didn't say anything about getting pregnant. It merely said that if you had a child under eighteen in your household for more than thirty days at a time you could be discharged. Your commander was required to discharge you. That's the way the regulation read and there was no legal way to challenge it. I actually started the process by talking with the director of Women Marines before I was discharged. I didn't want to get off active duty. I had almost ten years in. I had got married in 1967. This was in 1970. I was born in 1938.

"I am very familar with Jeanne Botwright's story because she came right behind me. We were both discharged just months apart. I went to the Board of Naval Corrections and challenged the regulation and, when she found out I had done it, she put in her request to do the same thing, so our careers kind of ran parallel. Jeanne was a former enlisted and she went through Parris Island. I think she was a drill instructor at one point. I know she was commissioned from the enlisted ranks.

"I did not go through Parris Island. I went through training at Quantico in '60 while I was in college. I had one semester left, at a school in Louisiana, and went back and completed that, got commissioned, and came on active duty in 1961, after graduating from North-western State College in business education and health and physical education.

"I was ordered to El Toro, California, where I served as executive officer of the Woman Marine Detachment and then transferred to the Recruit Depot at Parris Island in October of '64, which is where I met my husband, Bill.

"I had just reported in and was going to admin school when the colonel asked me to go sit in on a lecture this chaplain was going to be giving to the recruits. I went over and here was this tall, good-looking, tanned Navy guy in a white uniform and he gave a wonderful lecture to the recruits about morality and behavior.

"When I got back, the colonel, Julia Hamblet, asked me how it went and I said, 'Fine. He did a beautiful job,' and then I found out why he was there. It turned out they'd been having trouble with the chaplain who was normally assigned to the lectures, and Bill got called to fill in at the last minute because the guy couldn't be found.

"Bill was chaplain of the Second Recruit Training Battalion at the time. He had come on active duty in 1956 from a Baptist seminary in Louisville. He got sent to Vietnam in May of 1966 and we got married in May after he returned in April of 1967.

"I was at Parris Island three years and then, from 1967 to 1968, I served as the commanding officer of the Woman Marine Company, Headquarters Battalion, in Camp Pendleton, and went from there to the same job in Albany, Georgia, in 1969. So I had three different commands over the ten years before I was discharged, and in the course of that time I myself discharged many women Marines 'by reason of pregnancy.' It was painful for me and most likely for them as well because they were good Marines who wanted to stay with the Corps and they had to be separated. I was a major, the commanding officer of the women's unit at the Marine Corps Supply Center in Albany, when I was discharged.

"After I got pregnant and before I was separated from the Marine Corps, I wrote Col. Jeannette Sustad, the director of Women Marines, and told her I was not happy about having to get off active duty and out of the Corps, and she said, in essence, you have to because that's the way the regulation reads. So as soon as I was discharged in March of 1970, I started coordinating with her on the basic letter to Head-quarters Marine Corps requesting that the regulations be overturned on the grounds that the ruling was discriminatory: Men didn't have to be discharged when *their* wives got pregnant, but we did. My husband, who was in the Navy, didn't have to be discharged.

"I was in frequent contact with Colonel Sustad before I was discharged and she was talking to people at HQMC, more or less paving the way, so that when my letter arrived it didn't come as a big shock to them. She kept asking me to be 'patiently impatient just a little longer.' When she gave me the green light, I sent it forward and it was routed from one division to another, it went to JAG, the legal division, and it went to manpower. Finally, it was approved, and I was recommissioned as a reservist. I also gave birth to our son, Steven, on September 10.

"I went into a reserve unit and stayed as a Class III Reservist until 1976, when I came back on active duty. The Marine Corps had billets for sixty-six reserve officers. I applied for that program and came back on active duty. It was called Category VI. I was assigned as executive officer of the Reserve Liaison Unit at Camp Lejeune, North Carolina, and spent four years there. In 1976, as I said, I came back on active duty and then, in 1977, I submitted a request to the Naval Discharge Review Board to regain my Regular commission. That took forever. They had never had a request like that, and they didn't really know what the hell to do with it, to be honest with you.

"They called the people that worked individual cases recorders, and the recorder that got my case was evidently an alcoholic and he just drug his feet on it. Finally I went up there and pounded on some desks and went to the administrative part of the board and said, Look, I submitted a request years ago and I still don't have any action on it.

"I had a friend in the JAG office and I went and talked to him, and he rattled some chains up at the review board also. My request was finally approved by the Secretary of the Navy.

"Now, I was already on active duty, so all it took was changing my records from Reserve to Regular status, a little bitty piece of paper. Honest to God, it was a form called a DD215 and all it did was say the records should read as if I had never been separated. It had a great effect on your pension. When Jeanne got her notice back, after me, she retired in a month or two because she had enough time in. I did not. Besides, you know, I loved the Marine Corps. I had fought all this time to get the regulation changed, so why would I separate? So I stayed on active duty and I was subsequently transferred to HQMC and worked in the IG [Inspector General] Division, near the Pentagon. I was attached to Henderson Hall.

"It was a great feeling to think you had helped change the culture of the Marine Corps. For the longest time, whenever I'd go into a PX or someplace like that and in particular, after I retired, I would see a woman walking around in a maternity uniform, and I would think, God, look at that. Isn't that great? That kid is able to salvage her career.

"I didn't want to give it up. I had almost ten years in, and I loved the Marine Corps. It was my career, my profession. I had majored in education in college and probably would have done that if I had not come in the Marine Corps, but the Corps offered me a lot more than the field of education. That was why I fought so hard to get back in. I'm very gray-headed, and I say that one of the reasons is because of all those sleepless nights worrying about how I was going to approach this. You know, when you're coming up against an institution like the Marine Corps or the Pentagon, you've got to be dang sure that you got your ducks in a row, and that you're going about it in the right way. My other option was to sue, but my Marine Corps was not to be sued, especially when there was a way that I could hopefully change the regulations and the culture.

"I retired June 30, 1985, as a lieutenant colonel after twenty-four years, and went into real estate. I knew Doris Kleberger. She could work rings around you and me together. I've long been active in the Women Marines Association—I was the WMA representative to the Marine Corps Museum in Quantico, procuring uniforms and memorabilia —and I have to say the young women Marines of today don't show a great interest in those of us who went before. They don't know what I did, and I don't advertise it. I have a friend who served in Vietnam. She was my executive officer at Camp Pendleton. She was one of only a few women officers who went to Vietnam. She tried to talk about it to the girls at the WMA anniversary luncheon in February, but they didn't want to hear it. All they wanted to talk about was Iraq. They were not interested in the history of what women had done in the Marine Corps, and it just blows my mind. They don't really care, in my opinion. They don't know about regulations on pregnancy. They just take it for granted that's the way it's always been.

"I do want to say the Corps' response to women in the ranks has been very positive overall, and it's great that women have equality with the men. As for women participating in combat, as things stand I would rather see them in support, but if there were ever a necessity, then I would go for it."

CHAPTER TWENTY-ONE

JEANNE BOTWRIGHT

Lieutenant Colonel
1954–1957, 1958–1970, 1974–1980

I was one of those who challenged the pregnancy
rule. A couple of other women did too.

Jeanne Botwright was recommended to the drill
field as a corporal in 1959. She was called a platoon
sergeant, not a drill instructor. The primary emphasis for a woman recruit was "on being a lady," she
said, "and, yes, they had classes on how to apply
makeup." *(Photograph courtesy of Jeanne Botwright)*

Jeanne Botwright was in and out of the Marine Corps three times. She left the first time to get married and, when that didn't work out, she came back in, married again, then had to leave because she got pregnant. After following Mary Sue League in successfully challenging the regulation that required women to quit the Corps if they got pregnant, she rejoined as a reservist and eventually won restoration to rank as a regular. We spoke on the phone and it became apparent, as we talked, that giving women classes in deportment and personal grooming in the early days was not such a bad idea after all. She had served as a platoon sergeant in charge of recruits at Parris Island, and she knew what she was talking about.

"I was very patriotic. My father had been in the Navy and I had a brother in the Army and a brother in the Marine Corps. I graduated from high school in Erie, Pennsylvania, in June of 1952, and joined the Marine Corps in May of 1954. I contracted polio right after I graduated from high school. I wasn't able to work for a while and then when I recovered after a year and a half I became a clerk for an insurance company. I tried to go to college. I wanted to be a phys ed major but they wouldn't accept me because I'd had polio. So I thought, well, I'll show you, and I joined the Marine Corps.

"I took a train from Pittsburgh to Washington, D.C., and went from there to Yemassee, and got the bus to Parris Island. I arrived in the middle of an afternoon in May. I graduated from boot camp in August. I don't know if it was easy for all women but it wasn't hard for me because my parents were quite strict. 'Yes, sir' and 'No, sir' was common in my family, so I had no problem with discipline. My father was strict but he didn't abuse us. I went to aviation school in Jacksonville, Florida, then to Memphis for aviation electronics technician training. At the time I went there, women students were allowed to fly air-to-ground tests of the equipment they worked on. But then they took women off the program so I took an administrative drop because I didn't want to be in the program if I couldn't fly. I wound up in communications at Cherry Point.

"I got out as a sergeant after three years in 1957 and got married. It was one of those big mistakes, you know? It didn't last, so I got divorced, turned around, and went back in the Marine Corps as a Pfc. in August of '58. I went right back to Cherry Point and from there I was recommended to Parris Island for the drill field. I was a corporal by then. This was '59. There was no D.I. school for women then. As a matter of fact, they were not allowed to call us drill instructors. We were referred to as platoon sergeants, because only the men could be D.I.s. We didn't wear covers but I think we had red epaulettes on our uniforms. That, the rank on our uniforms, and our conduct were all that separated us from the recruits. That, and the fact that they traveled in formation and we did not!

"We still weren't allowed to fire the weapon or anything like that. The primary emphasis was on being a lady as well as being a woman Marine. They had classes on conduct. And, yes, they had classes on how to apply makeup. They had them in the enlisted program as well as in the officer program. They wanted to make sure you looked like a lady and acted like a lady. Some young women had to learn how to apply makeup. And there were a lot of girls who didn't care. That wasn't a primary concern of theirs. They joined the Marine Corps to serve their country, not to have somebody tell them how to wear makeup and how to do their hair. So that was a big change for them. Some resented it. You had to wear a certain shade of lipstick, your hair could only be cut a certain way, and worn a certain way, touch but not cover the collar. Now they can wear their hair long, as long as it's up when they're in uniform, as long as it's up and not hanging down the middle of their back. So many things have changed."

Carol Mutter, the first woman to attain the rank of three-star general in the Marine Corps, taught such classes to women officers at Quantico, Virginia. "If you look back at the 1960s and '70s," she recalled, "hardly anybody knew women who were in the military. It wasn't like today, when everyone knows someone who knows a woman in the service. Back then the assumption was that they were knuckle draggers and homosexuals, so the idea of the grooming classes was to help counter that perception. It was important that you understood how to wear the uniform and look good in it. The Marine Corps takes pride in men and women looking good. It was comparable to a young woman coming out of high school or even college and showing up for her first job not looking the way she should. They're still in a kind of teeny-bopper mode and wearing miniskirts. In those days, we had those."

Boot camp for women in the 1960s, Jeanne Botwright said, was not arduous. "We had exercise classes in the gymnasium, and we had field days, which would consist of volleyball, softball, and so forth, nothing like what they go through now. You did have to pass the swimming test, but they didn't have to wear packs or inflate the trousers for use as floats. That had not come into being yet.

"Recruits were trained by a senior woman staff NCO and two juniors. We didn't wear covers. I was a junior. One of the big things that came about when I was there was that you could not touch a recruit. If you wanted to correct something on her uniform that was improper, you had to say to her, 'I'm going to touch you,' or even, 'May I touch you?' You could not lay a hand on them or even imply that you were going to. This was 1959–60, all part of the fallout after Ribbon Creek.

"There was always a male D.I., usually a sergeant or a staff NCO, assigned to the women's platoons, and we never had a problem. He assisted with the drill portion and sometimes taught history and traditions from the male standpoint. They were thoroughly screened by the woman commanding officer. They were all married men, chosen as such. We didn't resent them. Women respond differently to men than they do to women. For example, when I was CO of a truck company in the Reserves I had only one problem in three years, because none of the men wanted to be disciplined by a woman. Conversely, women performed better for the male D.I. because they wanted to look good in his eyes.

"How did I come to challenge the pregnancy rule? First of all, I resented fact that I had almost sixteen years in and I had to get out. That came long after my term on the drill field. I got married again in

March of 1970, twelve years after my divorce. My second husband and I were both mustangs, officers who came from the ranks. I was a major, but I got pregnant and had to get out. There was no rule against being pregnant. You just weren't allowed to have dependents under the age of eighteen. Theoretically you could put the child up for adoption, and stay in. But I would never have done that. So I had to get out, in June of 1970. My husband was still in the Marine Corps. He was stationed at Camp Lejeune. He had been a first lieutenant but he reverted to staff sergeant because he didn't want to fly a desk—he wanted to be with the troops.

"In 1974, four years after my daughter was born, I got back in as a Class III Reservist. I was commanding officer of a truck company and went to summer camp with them and while at one of the summer camps the commanding general at Little Creek observed me with some of my troops and asked me if I wanted to go back on active duty as a reservist. They had a Category VI program where reserves could could fill a billet with a regular establishment, and that's what I did.

"The pregnancy rule had been an issue for a long time, and it had been challenged by a lot of women in all the services. Mary Sue League was the first to challenge it in the Marine Corps. Standing in our way was what we called the 'good old boy' network. Finally I got a letter from women officers at headquarters saying the Marine Corps was changing its policy to allow regular women Marines to stay on active duty if they had dependents. Since I was a reservist, I wanted to try to get my regular commission back so I could retire as a regular officer rather than a reserve. That way I would get paid sooner in my retirement. As a reservist, you didn't get paid till you were sixty-two, I think. As a regular, you got your retirement pay right away. So, in 1977, I wrote the Board of Correction at Naval Records, challenging the fact that I had been asked to leave when I had almost sixteen years in as a career officer.

"They rake you over the coals, drag up all your old records and all your old fitness reports and all your old conduct and proficiency marks, everything everybody ever said about you. It's a very lengthy process, a paper mill, but I had a lot of support. I guess it was all okay because in 1979 they restored me with the date and rank of my class.

That made me the senior lieutenant colonel in the Fourth Division Headquarters, eligible for retirement, which I took, on March 1, 1980.

"My second husband and I were divorced in 1990. We had two children. I'm very active in the Women Marines Association. We have a biennial convention and we have an interim board meeting and I stay busy with that. I'm national secretary of the Association. We have about 3,600 members. We try to perpetuate the tradition of women in the service, particularly in the Marine Corps, and we provide scholarships for dependents of Marines or for active duty Marines who qualify. We have a great volunteer organization: We volunteer at veterans' hospitals, children's homes, women's shelters, places like that. Until September 2004, Carol Mutter was national president. She is a fantastic lady, a credit to the Marine Corps.

"I'm 71 now. I live and travel full time in a motor home, a forty-foot RV. Home is where the jacks are down and the feet are up. My mail forwarding address is in Apache Junction, Arizona. I spend a couple months there, a couple of months in California, and the rest of the year I go around visiting family and attending association functions.

"I share the driving with another lady who was in the Marine Corps during the Korean conflict. We tow a car, and run right along at sixty-five miles an hour. We have a CD that tells you where all the RV parks are, what they cost, and so on. We look up the area where we're going, pick one out, and call for reservations. Then we show up, hook up, free the car, and go sightseeing. We've been to all the states.

"It was a good career. If I hadn't joined the Marine Corps, I probably would have stayed in Erie, Pennsylvania, the rest of my life. I had an opportunity to travel—I was stationed on both coasts, I was stationed in Hawaii, I went to Japan, Okinawa. I never would have done that if I hadn't joined the service. The evolution of women in the Corps, change for the better, has been extraordinary. You have to go with the flow, keep up with the times, and I think what they've done to attract women has been fantastic. I feel that I was born forty years too soon, and I so admire the women who are in now. I wish I could have done the things they're doing today. I will say they appreciate the accomplishments of the women who went before them. We've had a lot of them join the Women Marines Association and they always talk about how wonderful it was of the women who went before them to break the ice.

"We still haven't overcome all the obstacles. There are still a lot of men in the Marine Corps who oppose women belonging to organizations that work close to the front. For example, we have women truck drivers in Iraq and there was a big article about them driving through fire to resupply frontline units. A lot of men resented that. But attitudes are changing, slowly. I feel that if a woman can perform as well as a man in a job she should be allowed to hold the job, whether it's combat or anything else. Look at the women pilots. Some women in the Navy are flying combat aircraft. I feel if you're capable of doing the job, you should be allowed to do it. That's always been my feeling. That was my feeling when I was in. Why can't I do this? I'm capable. No, you're not allowed to.

"Every Marine, man or woman, truly is a rifleman nowadays. It was supposed to be true when I was in, but it wasn't. We weren't allowed to fire weapons, even though I did qualify, finally. I was the communications officer in Norfolk in the Vietnam era, in the '60s, and I had to qualify with a .45-caliber handgun. I got to be proficient with it, although I haven't fired it in a while. Do we carry one in the RV? Oh, absolutely. Not a .45, but we do carry a .38. You never know."

CHAPTER TWENTY-TWO

CHRISTINE HENNING

Staff Sergeant
1995–

When you go to D.I. school, they teach a basic cadence,
tell you to be loud and use your diaphragm instead of your throat.
Then you hear other people's cadence and you put your
own twist on it. I just kept practicing.

Christine Henning, staff sergeant. Her grandfather
was a Marine at Iwo Jima, although she did not find
out until after she had enlisted. She went from being
a series gunnery sergeant at the Parris Island Depot
to service in Iraq. *(Photograph courtesy of Christine
Henning)*

There is a cadence-calling and *a story-telling contest every year at the drill instructors' reunion at Parris Island. Staff Sgt. Henning took part in April 2004. I was impressed by her performance, and met with her later, after talking with Sergeant Major Denise Kreuser (Chapter 18), in Colonel Johnson's office. Months later, needing a photograph of her for this chapter, I contacted her mother, Jolene Henning, in Avoca, Iowa. She said her daughter was five feet four and 140 pounds, in her mid-twenties. "She was a very quiet girl, shy even," Mrs. Henning said, "and it was amazing she joined the Marine Corps. And then, when people here in town found out she was a drill instructor, they couldn't believe it. They said, 'How in the world did she do that?' I said, 'Oh, I've seen her in action. She likes counting cadence.' "*

"I graduated from high school in Avoca, Iowa, a town of 1,500, in May 1995," Chris Henning said. "I didn't want to go to college. I was sick of school. I wanted to serve my country. I wanted the challenge. I wanted something different. When I told everybody I wanted to join the Marine Corps, my friends all said I wouldn't make it. I'm a quiet person, if you know me as a regular person. So I had to prove it to them. But I would have a hard time getting yelled at because my parents—if I really got in trouble—they yelled at me, but I was a pretty good kid. My dad was a cop so I tried to keep out of trouble.

"My friends told me I shouldn't join the Marine Corps, that it would be too hard and I should join the Air Force or the Army, but I wasn't about that. My grandfather was a Marine at Iwo Jima. I didn't find that out until after I enlisted. His name was Curtiss Henning. He survived Iwo, and has since died. He never gave me no details about it. I just know he was there.

"The recruiters came to our high school, all the services. I just talked to the Army to get a feel for what they were saying, and the Air Force recruiter, well, I couldn't understand him. I didn't know what he was talking about. I just sat there and listened and didn't have a clue. Then the Marine Corps came and they just looked better, their presence, the way they carried themselves. The uniform. That's what started it.

"I signed up in August of 1994 but didn't leave for Parris Island till July of '95, after softball season. Boot camp was difficult. I wasn't used to getting up at zero 4 and zero 5, I wasn't used to marching around all day wearing boots. I was not an awesome PTer, so that was difficult for me. Back then you had to run three-quarters of a mile. We had fifty-two in the platoon. They broke us down and built us back up. Now recruits run from a mile and a half to three miles. The most we run is 3.2 and that's the motivational run at the end, the day before graduation.

"In the beginning we cried. We didn't want to be there. The drill instructors would come out and start screaming at us. We don't know what they're saying, we don't know what they want us to do. So that's stressful. We had to take tests and we helped each other. I remember that because I was scared: I was a bad test taker in high school so I was scared when it came time to take the tests. I thought I was going to fail. My bunkie helped me with some things I couldn't do. The recruits around you help you. I wasn't awesome at making racks. I wasn't an awesome recruit. I wasn't bad, I wasn't good. I was average. My bunkie helped me make my rack because I couldn't do the forty-five-degree angle thing on the corners.

"Now I can do it because I have to teach recruits to do it. Towards the end you build the teamwork, it all comes together. This doesn't happen at the beginning. You want it to start the first day you arrive but you know how it is for females all together in a squad bay: It's straight crabbiness. Especially when you're tired and hungry and some of them don't want to be there so they take it out on everybody else. But toward the end, we were pretty tight. We didn't want to leave each other.

"When I went through I had MCT, Marine Combat Training, here. Now they go to Camp Geiger for three or four weeks. They had that part of our Basic Warrior Training package here. After I left boot camp, I went straight to my MOS school. That happened to be Camp

Johnson at Camp Lejeune. I picked my MOS, Motor Transport, when I signed up in August of 1994. I wanted to drive trucks. I don't like to sit behind a desk. I like to be outside. It's fun to drive big trucks, especially when you get to drive through big mud puddles you wouldn't take your car through. I drove Humvees and Five Tons; they have bigger trucks they call LVSs—that's advanced training, because they're bigger. I didn't go to that. I drove Five Tons and Humvees for six years, and, when I'm done here in July, I'll go back to that.

"I volunteered for D.I. school, went there in April 2001, got here July 2001. It was like basic all over again, only harder. Every day I asked myself what the heck I was doing here, but I'm glad I went through. I love it. I'm going to miss it when I leave. I know I am. I enjoy counting cadence. When you go to D.I. school, they teach a basic cadence, tell you to be loud and use your diaphragm instead of your throat. Then you hear other people's cadence and you put your own twist on it. I just kept practicing. You just start out with, Left, Right, Left, Right, then you add your sing, so it brings a little motivation to the recruits, makes them want to drill. My recruits right now are in forming and all they get is Left, Right, Left—really slow. But, as they progress, they'll go faster and they'll get what we call singsong. Once you get to that point, you can change it up; some people do it to songs like the Marines Hymn or any song they hear on the radio, if they like that beat.

"I was a senior drill instructor the last cycle, and this cycle I'm a series gunnery sergeant. I have six D.I.s and just got a new one today, so now I have seven. They're in charge of my two platoons, 54 and 53. Women do everything the men do except pull-ups, although I had a recruit two cycles ago who did 21 dead-hang pulls. The last cycle I had one do ten or eleven. A lot of men can't do that. We do everything the males do except pull-ups and we have more time to do our running. We do push-ups, but not for any kind of score.

"We have two hours allotted for PT sessions and we do a three-, five-, six- and ten-mile hike. When we go to the Crucible, it's a six-mile hike out there, and a nine-mile hike back. That's considered PT. And McMap, the martial arts training, is considered PT, as is the Obstacle Course, the Confidence Course. We also do a circuit course with weights, curls, biceps, and triceps.

"We're only on the fourth day of training, and I had to teach a couple classes, but it's hard for me to go on their deck and not do anything, not interact with the recruits. I mean, I can, but that's not my job. It is real hard not to say anything because I know I shouldn't because of the billet I'm in. It's going to be hard throughout the cycle. Once we start training, I'll be in charge of PT so I'll do all that. The drill instructors will be there but I'm in charge. Also on the hikes. That's my only direct interaction with recruits. If I see something wrong, danger to a recruit or danger to a D.I., that's when I jump in. But one on one, I don't get that. I miss it already and I haven't even started yet.

"What do I like about being a drill instructor? I like being able to train civilians to become Marines. I like instilling discipline in them, showing them things about the Marine Corps, teaching them about the Marine Corps. Some of them come here because they have nowhere to go in their life. Some come here because they want to fix things in their life. And it's a pride thing. When you give that recruit, who has nowhere to go in her life, some place to go, a goal she's going to meet, you did it, you trained her to do that, to become a Marine.

"I'd probably say half come to escape from something. I don't know the specifics but a lot of recruits do come here trying to get away from something. It may be family, limited prospects, a small town, no jobs, drugs, alcohol, abuse. A lot are trying to get away from abuse, verbal, physical, or sexual. Some just come for education. Some come for travel."

Asked what she would do when her tour on the drill field ended, Staff Sgt. Henning replied, "I'll probably stay in. I reenlisted in 2002, to 2006, and I'll be in eleven years and two months by then.

"Do I want to go to Iraq? Yes, sir, I think that's why I joined the Marine Corps, to serve my country. If I don't want to be there, I shouldn't be in the Marine Corps. I know when I go to my next unit in September I'll be in that next group that goes, or the one after that. I expect to go to Iraq, especially in my MOS, Motor Transport. The idea is to give the ones that are over there a break. I have a friend who came back in July and now he's back over there again, nine months later. I want to do my share.

"I'll be there but don't know where I'll be. I don't think anybody wants to go into combat. But I'm going to do what I have to do, do what the mission requires. So if I have to go, I have to go. But nobody wants to. Nobody wants to stand up and shoot somebody. I'm just going do whatever the mission is. I'm going to accomplish the mission.

"I didn't handle guns before I came into the Marine Corps. I'd shot my dad's shotgun once and maybe a B.B. gun. I like to go to the rifle range. I shoot well. I was one point away from Sharpshooter when I was in recruit training. I shot a 209. You got to have 210 to get Sharpshooter. There's three categories, Marksman, then Sharpshooter, then Expert. I'm an Expert now with the M-16.

"If somebody comes up and I have to shoot him I'm going to shoot him. Whether I'm in the front line or not. I'll do what I have to do. Might not be on front line, but I'll shoot him if I have to. If they can get in our rear, then we're on the front line because they will have come into our area."

This interview was conducted at Parris Island in April 2004. A little over a year later, on May 2, 2005, the following message could be heard on Staff Sgt. Henning's cell phone: "Hey, it's me. I'm in Iraq, uh, give me a call back in . . . I don't know, November or October time frame, and I should be able to give you a call back. Otherwise I'll call ya when I can call ya. Bye."

POSTSCRIPT

Reached in Iraq in late September, after she had been there six months, Staff Sgt. Henning said by e-mail that she was stationed at Camp Taqaddum, serving as Motor Transport Chief of Base Operations. "The living conditions are not too bad," she wrote. "We are living in wooden buildings or tents. There are six females in my wooden building. We have real racks and a full-size fridge. You have to find them or buy one yourself. We have a TV and a DVD player.

"Things are going pretty good. I have five more months to go. It is very hot here in the summer and I am about to see what it is like in the winter. I am not driving big trucks. I drive once in a while but we don't go outside the wire. We transport supplies or move generators, wire, and other things to various sites around the base. We keep really busy.

"Some of the smaller bases don't have it as good as we do. I work in a hard-stand building left over from Saddam's time. We have Internet capabilities, a weight room, cardio room, game room, movie tent, and an Internet café which also has regular phones. There is also an AT&T phone center. We have a fairly nice PX like a 7-Eleven, with electronics and clothing.

"The food isn't too bad; it just repeats itself a lot but all in all it isn't too bad, a lot better than a MRE [Meals Ready to Eat—in the field] three times a day. I guess I can't complain about that. The PX also has some food, chips, tuna, and juices you can carry if you are out on a job and don't make it to chow or you are going on a convoy.

"As I said, I live with five other females and we all have different jobs here on the base. We don't really see each other that much but, when we do, we don't talk about the war. We try to talk about other things when we are sitting in our hooch watching TV. So I don't know what they think.

"I don't really know what I think either but I am glad that I am part of the war trying to help the Iraqi people have a better life. Even though I don't go outside the wire, the job I do here on the base helps the ones that do have to go outside the wire. I am glad that I am able to contribute to the war

"Iraq is not a safe place. Anywhere can be dangerous. I myself have not encountered any IEDs, but some of my friends have. The only dangers I face are the incomings. Actually, while I am writing this e-mail, I am in Code Red. We just got hit. That is the only real big danger. The thing about them is you don't know when they are coming, so you always have to be prepared. A lot of times they hit when the sun is coming up and when it is going down. Some have been a little closer than we would like but fortunately we have been pretty lucky and haven't had any one get hurt or killed."

Staff Sgt. Henning said she expected to return to the United States in February or March of 2006 and she signed off with these words: *"Pain is weakness leaving the body. Pain is weakness of the mind, body and soul."*

Counting Cadence

Cadence counting has been part of the military from time immemorial. It is a singsong rhythm designed to help marching soldiers look sharp and keep in step. Rick Arndt, who was sergeant major at the Recruit Depot at Parris Island, remembered how a master sergeant at MCRD San Diego taught him to call cadence. "He chewed my ass one day when I was a sergeant on my first platoon. He said, 'You know what, Arndt? You need some help.' Well, I'm not going to tell a master sergeant that he's full of shit. So he took me under his wing, he'd walk with me and he taught me how to call good cadence. You use the same rhythm as your windshield wipers, 120 steps per minute. As long as you can dance, you can call good cadence; if you can't dance, you can't call good cadence, and it's been proven. You can watch a Marine at an event and, if that Marine can dance, he can call some cadence. Or if he sings in a church: I worked with a Puerto Rican who's now a sergeant major, Jose Raphael Nazario. He learned to sing in church. When he went through recruit training, he could not speak good English, but after he learned it, he could sing and call good cadence. We all got around him and learned.

"Good cadence will motivate the hell out of some recruits. You talk about a recruit standing tall and leaning back and picking up his feet and putting down his damn heels. Good singsong cadence will make or break some recruits. It's the same with running. If you put somebody out there with a platoon who can't call cadence and has no rhythm, that platoon will fall apart and they'll start sucking wind, and it's just a nasty-ass run.

"Staff Sgt. Rory was one of my platoon sergeants at Camp Lejeune when I was the company gunny there. The MCCRES Hump is a twenty-five-mile hike with a forty-pound pack and you have to do it in less than eight hours. That's a MCCRES standard. Well, when you get to the twenty-two-mile mark, you're gone, your feet are gone, your legs are gone, you're cramped. Everything's gone. And Staff Sgt. Rory started calling marching cadence while we were walking down the road coming up River Road at Camp Lejeune, and I'll tell you what: Everybody got their distance, everybody covered down. It took about half a mile to get people to start sounding off, but they all did it and everybody walked across the damn line when the clock was ticking, all because that staff sergeant had enough in him to be able to call the cadence. He could have been like a lot of other people, just worrying about himself and thinking about how miserable he was, but instead he helped everybody there make it across with his calling cadence. It takes your mind off your misery; it gives you some soul, gives you something to think about. It's amazing. I don't even know how the hell to describe it, but it works."

Women Counting Cadence

Lefty, right, a lo right
A Lo right alaya
We love to double-time

Navy, Navy, I'm in doubt
Why your bellies are sticking out
Is it whiskey or is it wine
Or is it lack of PT time?

Navy, Navy, where you at
Come on out and lose the fat
Don't be scared and don't be blue
Popeye was a sissy too

Moto, moto, got a lot of motivation
Dedi, dedi, got a lot of dedication

Motivation, Dedication
To the Corps, our Corps, your Corps,
   my Corps, Marine Corps

oorah oorah

PT, PT every day
Builds your bodies the Marine Corps way
Oh, yeah, here we go, pick it up, out front
My Aunt Molly was a WM
She liked whiskey and she liked men
She might be dumb but she ain't no fool
She never came to a D.I. school

Parris Island, how you doing?
We're doing fine now
Mind over matter
If you don't mind
It really doesn't matter

Pain in the feet
Pain in the knees
Pain in the legs
Get some, I love some

Hey, commander, can't you see?
A little PT is nothing to me

I used to sit at home all day
I let my life just waste away
And then one day a man in blue

Said, Girl, I've got the job for you
A free room and board with paid-up boots
And a brand-new tailored business suit

There's travel and adventure and loads of fun
You'll even learn to shoot a gun
Free medical and dental benefits
And lightweight aluminum campaign kits

I'm glad he convinced me to keep your body tight
Eight hours' sleep each night
Believe me, girl, he said with a grin,
Sign these papers and I'll get you right in
And I got on the bus he winked with his eye
I never knew a man could be so sly

I'm not complaining, don't get me wrong
I've enjoyed myself, growing big and strong
But two things the recruiter didn't promise me
A world war starting
And a cup of tea.

Oh, yeah, here we go

PART SIX

THE NEW BREED

The More Things Change, the More They Stay the Same

These recruits are entrusted to my care. I will train them to the best of my ability. I will develop them into smartly disciplined, physically fit, basically trained Marines, thoroughly indoctrinated in love of the Corps and country. I will demand of them, and demonstrate by my own personal example, the highest standards of personal conduct, morality and professional skill.

—The drill instructor's creed

MARINES TODAY appear to have come a long way from the group that inspired the following observation by Eleanor Roosevelt, the wife of President Franklin Delano Roosevelt, in 1945: "The Marines I have seen around the world have the cleanest bodies, the filthiest minds, the highest morale, and the lowest morals of any group of animals I have ever seen. Thank God for the United States Marine Corps!"

There are no jailbirds today, recruits can't have too many tattoos, they have to be high school graduates, they get to apply Skin So Soft to ward off the infamous sand fleas, they are encouraged to drink lots of water, and they get to run in shoes made for the purpose. Old-timers such as Mike Mervosh are annoyed by the way decorations are handed out so freely, and some think the old, harsher ways were better, men were tougher, and women were nowhere near the combat zone.

Yet, the more things change, the more things stay the same. Have the changes been for the better? The following six chapters, featuring a drill instructor from the Recruit Depot at San Diego, four sergeants major, and the head of the drill instructor school at Parris Island, may provide the answer. The old days are gone; the new days are here; the Corps marches on.

There was certainly no question about the Leatherneck performance in Iraq: Marines and the Third Infantry Division of the Army made up the leading forces in the drive toward Baghdad in 2003. In addition, Marines had the challenge of pacifying Anbar province, containing Fallujah, Ramadi, and Haditha, sites of some of the worst insurgency in 2004 and 2005. Marines lost 530 of the more than 1,820 U.S. personnel who had died in Iraq by late August of 2005, according to the Associated Press. Roughly 23,000 Marines out of a total of 138,000 members of the armed forces were stationed in Iraq, which meant the Marines made up 17 percent of the force, but they took about 30 percent of the casualties, primarily because they were given some of the toughest jobs to deal with, such as Fallujah and Ramadi.

CHAPTER TWENTY-THREE

RODOLFO RODRIGUEZ

Staff Sergeant
Marine Corps Recruit Depot San Diego
1995–

If you're the kind of person that is okay
with mediocre performance, then this is probably
not the place for you.

A native Californian, Rodolfo Rodriguez was drawn
to enlist by watching movies such as *Heartbreak
Ridge*. "I saw that uniform," he said. "I saw what the
Marines were." He became a drill instructor at the
Recruit Depot in San Diego in 2003. (*Photograph by
Jean Fujisaki*)

I met with Staff Sgt. Rodriguez during a visit to the Recruit Depot in San Diego. He was responsive, courteous, and very proper, extremely correct in his answers to my questions. He seemed reluctant to acknowledge that recruits almost invariably arrive after dark. His commitment to his job as a drill instructor was palpable.

"What do I do to correct a recruit's behavior? You can't punch them or swear at them. I don't think that's effective, either one of those two tools. But I'll tell you what is effective, and that's documenting. I personally will always document every recruit's behavior. That's the right thing always. I've learned that always comes in handy. I click on his name and put in a comment: 'So-named recruit disobeyed an order. When told to go directly to the head to fetch cleaning solution and swab the quarterdeck, he decided to do something else.' So he wasn't following orders. I document everything, so their performance is like a series of tests. If you have any questions on a recruit, it's all documented.

"You let the recruit know about it, and that's very effective because he knows now that it's official. There's only so many times that you can get off line and mess up. So that's one.

"Two, you can award them IT, Incentive Training." "Incentive" in this context means PT for failure to follow orders. "You have a card with certain authorized exercises. It lists how many you can do, repetitions, with breaks in between. It allows a good five or six minutes and a combination of exercises which gets the point across. There's a way you do push-ups on a three-count and you can order twenty-five of them. So you go 1-2-3, that's one. And you do twenty-five of them. You can say arm rotations, and then after three minutes, you give 'em a break for sixty seconds, thirty seconds, whatever the IT card says. That is actually one thing you have to have in your hand. When you IT a recruit, you've gotta have a watch with a second hand and an IT card, and you'd better know where you're at on that IT card. There's no duckwalking, no running around the perimeter. If that happens, it's gonna get noticed.

"There's a lot of supervision out here. I mean, my career would never be worth that recruit that I'm trying to get a point across to and miss one of these steps. I will have an IT card; I'm not trying to memorize it, because I'm looking right at it, and you better know where you're at on that IT card, because someone's watching."

For Marines such as Rodriguez, the job holds great appeal. "I put in a request to be considered for D.I. school a couple of times and then I got orders in October 2002, arrived here in January of 2003. It took almost three months to go through. D.I. school was the most difficult thing I have ever done, up to that point, the same way I thought when I first went through boot camp. It was constant and it went right back to what I had considered the most difficult part of boot camp: the constant attention to detail, the big demand for perfection, learning the Standard Operating Procedure, learning close order drill, all the key areas that encompass being a drill instructor. Three months is not a very long time to learn this, and they wanted perfection, and you had to perform. It was very demanding. Yet things I considered almost to be impossible at a certain point I would later find myself doing. So they were definitely correct, and I graduated. Now I think training recruits is far and away the most difficult thing I've ever done in my life.

"I'm twenty-eight years old. I was born on May 21, 1976, and raised in Los Angeles, and I graduated from high school in Fontana in San Bernardino County. My wife, Erika, and I have two boys, Rudy Jr., nine, and Antonio, who turned one on November 26, 2004. I wanted to be in the Marine Corps from early on, when I started watching movies like *Heartbreak Ridge*. I saw that uniform, I saw what the Marines were, that look, that was for me. Vietnam was long over, but there were Marines in Beirut, Lebanon, there were Mideast issues even then.

"I can't put my finger on what it was exactly. I knew I liked them and it wasn't unusual for young men from the area I was in to graduate and move on to the Marine Corps. I must have known six or seven guys who thought it was a good way to go. I joined January 24, 1995.

"I got here very early morning. It was still dark. It was raining. I bused in. I was picked up in San Diego, kept overnight, and brought in early in the morning, in the dark and the rain.

"I thought boot camp was the most difficult thing I had ever done in my life. I always lived my life structured. I had pretty strict parents but at that age, when I think back now, it was just the shock of being away from home, being away from everyone you know, your comfort zone. That's shock enough. That takes adjusting. But you get over that in a few weeks. The other part wasn't so bad. I was pretty physically fit so I actually enjoyed it. I'm six four now, weigh 200 pounds, although I wasn't that big then.

"What was it like, getting yelled at all the time? It was just the non-stop training demands, go, go, constant correction from the drill instructors. I never had so much attention in my life. That's the best way to put it. Perfection was never a big thing in my life. I had lived eighteen years getting by on the bare minimum, and, when I got here, they demanded and demanded and they didn't stop.

"I graduated April 14, and ten days later I went to Pendleton, to Marine Combat Training. I'm not Infantry, but I had to go to school to MCT, which is just more combat training we all have to go through if we're not Infantry. You go there to further your infantry skills, because every Marine is a basic rifleman. I'm an 04-11—Maintenance Analyst. Basically I keep track of all ground maintenance in a unit, everything that goes into maintenance or requires preventive maintenance, tracking parts and so on, anything ground-related, that doesn't fly, except for parachutes. It could be motor vehicles, weapons systems, missiles, communications, even dive gear.

"After MCT, I was with the Second Battalion, Fourth Marines, 5th Marine Regiment, at Camp Pendleton. I've been on deployments. I've been to Okinawa, I went on ship to the Persian Gulf and to Australia, Thailand, Africa, Pakistan, Afghanistan, Singapore, and Bahrain. I've covered a few miles. I loved it. It was very interesting to me, an experience I would never have been able to get anywhere else. I was representing the unit maintenance-wise on these deployments.

"Once I got into D.I. school, we went over the history of recruit training, learning about the McKeon incident and other things that all tie in with the Standard Operating Procedure. You're tested on the

SOP every day. You cannot graduate unless you pass it. We're tested on it every three months, every time we pick up a platoon. The SOP is very straightforward. I have never come across a question I did not find the answer to in there. Basically, it's the rules. It tells you what you can and can't do. In the sense of doing this job, we test each other on it all the time. The moment we see someone violating the SOP, even though he may not know it, we fix it right away.

"For example, I don't know that you were ever 'allowed' to swear at a recruit but I do know the SOP says foul language is greatly discouraged. I think what they mean by that is, you can't say nothing towards a recruit like—pardon me—'You dickhead!' You cannot say that. That is absolutely wrong. You are in violation of the SOP and you will be punished for it. But if you were walking and you were trying to loosen up something that's tight and you say, 'Shit, it's tight,' that also is discouraged. We would like to think you could use a better choice of words to express yourself, but you didn't mean that towards any one. That's just common use of an expletive. We discourage that greatly and that's something that a group leader will come up and say, 'You need to watch your language; that's unprofessional.' Because it is.

"There has been a lot of evolution, in recruits as well as training. I've heard about recruits going into the Marine Corps from jail, but it's not like that now. You've to be a high school graduate. I never personally met anyone who came in on a GED, so I think just about everyone these days has a high school diploma.

"It is not easy to be given the great, awesome opportunity of standing on those Yellow Footprints. It is a challenge just to get to that point. You have got to be able to pass written tests and, you know, drugs, you can't be a drug user. You have to be clean, your weight and height standards have to be met, and you have to be able to do a minimal physical they call an IST, an Initial Strength Test, a mile-and-a-half run, two pull-ups, and some crunches.

"Is it typical to deliver recruits in the dark to enhance the sense of disorientation and transition from a comfortable past to a new and uncertain future? I couldn't actually . . . I don't know. I can't say that's true because they don't know where they're at anyway. So I can't say. I would probably also say there's certain other factors, logistical factors, entering into it. We have a schedule we have to maintain and I think we just like to get an early start on things, because we have a full plate of things to do. I'm not a recruiter and don't work in Receiving so I've never really witnessed it, aside from when I went through. I can tell you that's something I never heard—that we come up with a plan to get them in here in the dark. That may be, but I've never done Receiving. I'm not part of Receiving Company. I've only been on the other end in the Training Company.

"After they get off the bus, they're processed for a week. They have to pass medical and dental exams, get their service books put together. I do know none of the standards have changed much in the last ten years. But if a recruit has lied or done anything he shouldn't have, he gets a chance in what is called Moments of Truth, when he's allowed to come clean about things he might not have admitted before. Just being separated from their families at such a young age is culture shock enough. Anyway, if they get here Monday or Tuesday you'll get them from Receiving on that Friday. The Receiving Company drill instructor will deliver them to your squad bay and he'll give them a head call and then sit them down. They will have dropped off all their gear on their side racks earlier in the day. The Receiving drill instructor brings them back after they have had noon chow and sits them in a kind of classroom formation.

"Then the drill instructors will come out, we'll be introduced, stand up there, the series gunnery sergeant will do the oath, I mean the drill sergeant's creed, not an oath, I'm sorry. We'll go through that. Some companies don't. We do it in my company. The senior D.I. will give them a speech, go over lot of things, let them know who he is, who we are, what he's there to do for them, what he expects of them, who the recruits can go to if they feel they have an issue, that they've been violated. He's the one they can go to. So they're given that. They're told smoking is absolutely not allowed.

"The senior is the most experienced drill instructor. He wears a Black Belt. He holds the billet of platoon commander. Then there's Green Belt drill instructors. So you have three in charge of the platoon. There's really no pecking order with the Green Belt drill instructors, like, I'll be the first one in charge and you're the second but there is an experience level. If I've done three cycles, and he's only done one, and he just got here, obviously the burden of experience would lie on myself. If one of us has a little more experience than the other, we would try to use that to the platoon's benefit and that one would make sure to pass on his wealth of knowledge to the others. But there is no pecking order. Most commonly you have two Green Belts and a Black Belt to a platoon. We do refer to each other as Hats, but it's not something we use loosely in front of recruits. We don't have the J, the heavy, or the third Hat. We're not allowed to smoke around recruits either.

"After the senior drill instructor welcomes the recruits, he will turn the meeting over to the D.I.s. The one with the most experience will let them know that at no time while on the Depot will they do things such as run in flipflop shoes, run if the ground is wet, run up and down ladder rungs, extend or force your body outside of portholes. If you get lost, he tells you what to do, how to act if you see a recruit get injured, and so on. Then, if they're split in two groups, he'll stand one group up to make sure their legs don't fall asleep and tell them to get on line and stand in front of their gear. Once they're out of the way, he'll get the other group up and make sure their legs are not asleep and they're not falling over, cracking their skull or something, and he'll get them on line. From there to the next two days it's in-house procedures, teaching them how to conduct themselves in the squad bay, how to walk, talk, the whole nine yards.

"We're very conscious right now that we're training recruits who may go to Iraq. I've already had recruits go there. We're acutely aware that some of them might not come back. There is no time for misconduct. We have a mission, and that mission we have to accomplish. It is a serious job. They have to be trained right. They have to be basically trained and indoctrinated in the Marine Corps before they go on to their specialized infantry training or their MOS schools.

"No job I have ever done in my life comes close to being a drill instructor. I finished with my first platoon in April of 2003. It was a good feeling just to be done. It was a great feeling. The standard tour for D.I.s is three years. Mine will expire in March 2006. I put in 110 to 120 hours with a recruit platoon over the seven days of the week.

There is someone with the platoon at all times. I'm with them fifteen hours a day. We give them everything we have, and we have to pull from deep down inside. Like, I know it's only 3:00 P.M and I still got another ten hours to give them. Where do I pull that out of? I pull it out, still give them more. Because they came here for this, they deserve it. But even if there was no war, we'd still give 'em the same way. The war going on right now doesn't really affect us, because this is the way we always train them. This is what's expected of us, to give those recruits that came here all we got. They deserve it. They volunteered for this. They came here. We have to give it.

"At the end of every day out on this job, when I go home, my wife takes a look at me and she's taking pictures of me with a digital camera and I see them later on and I go, 'Oh, someone's tired.' But with the recruits, it's 110 percent business, energy, full throttle.

"The recruits will feed off you. If you move slow and you look tired, they look tired. I can tell you that. Although we are given every opportunity to eat and sleep and we are taken care of as drill instructors, there is no way that I will eat until every single one of my recruits has eaten. There is no way that I will go to sleep until my recruits are in bed. Those are things that just don't happen. If I have a problem, I start coming apart inside. That's just that Marine Corps leadership. You know there's stuff you gotta get done. There's a lot of moving fast in order to meet that next evolution of the training schedule. It takes a solid team and a solid group of Marines. It's a lot of hard work. If you're the kind of person that is okay with mediocre performance, then this is probably not the place for you.

"I've had guys go UA (unauthorized absence) just like they do at Parris Island, but they don't get much in the way of a head start and they're usually caught on the premises. The reason for this is four recruits stand fire watch throughout the night, one at the front of the squad bay, where the eighty-plus recruits live and sleep in double racks of bunks. A second recruit stands at the back and the other two wander around to make sure nothing unsafe is happening, no stealing, no food around or anything, no fires being started.

"Fire watch is an old term. It's mostly a security detail. The rifles are double-locked right behind the bunks. The recruits on watch rotate every hour or so and, once they rotate, part of their duty is to count everybody, count all the weapons. And if one comes up short, they notify me right away. So, say they counted recruits at 2200 and somebody goes UA between that hour and 2300, when they count again. The drill instructor sleeps at night in the duty hut in the squad bay. It's a small little office with a big window you can see out of but they can't see in. I'm in the duty hut, and I'm told within an hour if a recruit's missing. I do a check, he's not there, I call PMO, the Provost Marshal's Office, and, in my last case, they found him later on by the base gas station.

"I think there's a perception out there that people think that Marines are brainwashed like robots, but in the Marine Corps we put a lot of emphasis on small-unit leadership. For example, we have billets, or ranks, in recruit training. Some recruits are squad leaders or guides. Those are leadership billets. It was never simply instant obedience to orders. I never experienced that. When I came to recruit training, I was told this is how we do things, this is how you will do things, this is the way it is. You learn the ways and you go through. There's an old saying that, if there's two Pfc.s, two private first classes, or two privates, one is in charge because one of 'em has been in longer. There's that leadership."

CHAPTER TWENTY-FOUR

CLINT KREUSER

Sergeant Major
Iwo Jima Specialist
1982–2004

We've been practicing [urban assault]
around the country in Cleveland and L.A. and some of the
rundown parts of some of the major cities. Helicopter assault on the rooftops,
taking the buildings, street fighting—all that stuff.

Sgt. Major Clint Kreuser and his wife, Sgt. Major Denise
Kreuser, on the summit of Mount Suribachi, where the
American flag was raised during the battle of Iwo Jima,
February–March 1945. "You'd be a liar if you said you
didn't feel something," Clint reported. "I stood there
and I got a chill. It's a pilgrimage you have to make if
you get the chance." The island was returned to the
Japanese in June of 1968, during the Johnson administration, and a lot of old Marines are still unhappy about
it. *(Photograph courtesy of Clint and Denise Kreuser)*

I spoke with Sergeant Major Kreuser at Parris Island in April 2004, a few months before he retired, in July of that year. His wife, Sergeant Major Denise Kreuser (Chapter 18), replaced him as sergeant major of the Support Battalion. "It was big joke down there," he said more than a year later, "because they didn't need a lot of turnover. They were saying now I was retired she was just going to call me on the phone all the time, to ask me this or ask me that. Actually, it was nice that she came in and I didn't have to worry about turning something over to somebody who was just getting there." Clint now concentrates on his business of selling ribbon and medal displays on the Internet.

"I'm sergeant major of the Support Battalion, which provides for the needs of all four training battalions here at Parris Island. We rehab medically injured recruits and recruits who are physically incapable when they get here. We also teach combat swimming, the academics, and martial arts. We have one company that gets them off the bus and processes them through their first week here, plus we discharge any recruits that don't make it. So there are a lot of moving parts.

"The wash-out rate is 9 to 10 percent for men and about 12 percent for women. The rate has held steady over the years, even though we have a generally higher-quality recruit today. For example, you have to be a high school graduate to get in. We do get some who earn GEDs, but it's an extremely small percentage. As the recruits get smarter—computer wise or street savvy—we adjust training to fit them. We make it more challenging mentally.

"It's no secret that years ago, in the fifties, sixties, and seventies, recruits were hit a lot. The thought back then was they had to hit people to get through to some of these kids. Nowadays recruits are a little bit smarter. Hitting them is not necessary. It always was prohibited— I have an SOP from 1943, during World War II, and it states that it is not acceptable to hit a recruit—but it happened and people turned blind eyes to it. We deal with it severely when Marines get caught doing it nowadays. But it's very rare that it happens.

"I joined February 25, 1982, from Kenosha, Wisconsin. I graduated from high school in 1979, went to college for a year. I was in Army ROTC, but that wasn't moving fast enough, and I wanted to do something now, so I went and saw a recruiter after that first year and one thing led to another. I had always thought about the Marines.

"I did my recruit training at Parris Island. People from Wisconsin and part of Michigan are supposed to train in San Diego, while the rest of the country is pretty much split by the Mississippi: Those from the east side go to Parris Island. My recruiter was mad I got to go to Parris Island instead of San Diego, but all I ever heard about was Parris Island, so I wanted to come here. It's hard to explain. All the movies you see are Parris Island. The movie *The D.I.*, with Jack Webb, was about Parris Island. You never hear about San Diego. I wanted to come here.

"I finished boot camp in May of '82, went up to Camp Lejeune for two months for supply school, then to Okinawa for a year. I liked it a lot. I worked in the office with a Marine who had a car, which was very rare back then. I went to the PX and bought a camera, and every weekend we'd jump in his car and travel around. I took hundreds of pictures, lots of battle sites. Okinawa was the scene of some of the worst combat in World War II. I stood on the spot atop the Maeda Escarpment where the conscientious objector Desmond Doss, an Army medic, lowered fifty to a hundred injured soldiers, for which he received the Medal of Honor. They give tours now, and take you around and explain it. The First and Sixth Marine Divisions fought there as well.

"I came back to Camp Lejeune and served in an artillery unit for fifteen months. A gunny kept pushing me until I finally volunteered to go to Parris Island as a drill instructor, in January of 1985 to March of 1987. My second tour was from July 1990 to September of 1992. It has been three-year tours since 1996.

"After the second tour I went back to Lejeune as the battalion supply chief of the Second Combat Engineers until 1995, when I requested orders to go to the staff NCO academy as an instructor. I taught there until July of 1997. My wife and I went back to Okinawa in 1997.

"We had met in '85 in D.I. school. We were in the third squad together. I sat right behind her. Four of us studied together all the time. We'd get something to eat after school every night. We took our binders with us and just studied all the time. After school we dated for two years, the whole time we were down here, and then ended up getting married at the justice of the peace office in Beaufort in December of '86, three months before we left.

"Female boot camp was not as long as the males' back then because women didn't qualify with the rifle. They were two weeks shorter than us. Platoons never matched up, so when I was off, she was still working and when I was working, she was off. And you'd have duty every third night. It was hard to date.

"Females finally started going out to the rifle range that December, so their boot camp was lengthened by two weeks, and finally we each had a platoon that matched the other—on mess and maintenance. When the recruits were done with the rifle range back then, they would do mess duty in the chow halls, and the drill instructors would have a little time off. We had three or four days off together so we ran downtown, got the license, got married, went down to Disney World for three days. We came back Saturday, rushed back to work Sunday, and finished our platoons out. I was twenty-five; she would have been twenty-six.

"They do have base housing but we live off base, in Beaufort, ten miles away. We bought a house. I'm retiring this October. Denise is retiring next June 2005 [she retired in March of '05]. She will have twenty-six years and eight months in. It'll be twenty-two years, eight months, and six days for me. We have a Web site business, selling military medals and ribbons on the Internet. We do shadow boxes for retirees and vets. We're going move to Jacksonville, Florida, buy ten acres and build a house. We like warmer weather. Up in Wisconsin it's below zero in January, and she said she didn't want to be eighty years old and have to go out and shovel snow. And I said, Well, I don't want to be eighty years old and have to tell you to go out and shovel snow. So we decided to live down south.

"The stresses on a married couple in a Marine Corps setting are a lot less than you would think. I was talking to a first sergeant at the club and he was saying how nice it must be to be able to talk to your spouse about problems at work and have her know what you're talking about. I said that is one of the reasons we never have an argument. We never fight about anything. When one of us is tired, the other one knows why, and when one of us wants to go off and vent about what happened at work that day, the other one understands.

"The sergeant major runs things in the Marine Corps. There's a sergeant major in every unit from battalion and above. There is one sergeant major for every battalion, regiment, and division, only one, and he works directly for that commander. Rick Arndt, the Depot sergeant major at Parris Island [in 2004], works directly for the commanding general. The technical definition is senior enlisted adviser to the commanding officer. We advise him on anything and everything that pertains to enlisted matters, which is about 90 percent of everything that goes on in a unit, whether it's morale, leave, pay, discipline, or punishment. Anything. If he comes up with something, says he wants to start this program or he thinks the unit should start training in this fashion, he'll bounce it against his sergeant major first, and get input from our experience. We let him know the pros and cons of it. The CO doesn't always do what we advise him to do, but he doesn't have to; he's the one who makes the decisions. But they all understand that when we're behind closed doors I have just as much say as he does. I mean I've yelled at my CO, he's yelled at me. We've had it out. It's a free exchange back and forth. But when the door opens up and I walk out, he's the boss and I'm on board with his decision, whether it's along the lines that I advised or not. I go out and I enforce the policy, pass the word, and that's what we're doing.

"I go out during the day and watch the drill instructors in the Fleet. I watch my Marines, talk to them to see how they're doing, how they feel, their morale, their personal problems. Is something interfering with their work? And I come back and tell the CO. I am supposed to keep the whole machine well oiled and working.

"Sergeant major is the top NCO job in the Marine Corps. There are nine enlisted ranks. The rank of sergeant starts at E5 and goes up from there to staff sergeant, gunnery sergeant, first sergeant, then sergeant major is E9. Gunny is an important job; he is usually the most experienced one, at that level. Down at the company level, the first sergeant performs like a sergeant major.

"There are only about 440 sergeants major in the entire Marine Corps, which itself numbers 175,000. At our peak in World War II, we had 485,000 Marines. But the sergeants are a small community and we all know somebody who's been to Iraq and they've all said the quality of the recruits coming out of here and going over there is superb. The First Marine Division commanding general also sent a very positive message back to the Depot. We haven't modified training for Iraq. That would happen more in the infantry school at Camp Lejeune. Here we concentrate basically on making Marines, with focus on the mental and physical aspects. Everything else gets done at Lejeune.

"That said, there is a very heightened awareness at P.I. that we are training Marines who are going to Iraq and maybe not coming back. There might even be a subconscious thing with the drill instructors tending to look at the recruits a little bit more intensely to be certain they've got what it takes. We'd break them here—that sounds like a severe word—we're not trying break them physically but mentally, to be sure they have what it's going to take. Because some of them would rather quit at the drop of a hat when things get tough, and we'd rather have them do it here and send them home than have them do it over there. The Marine Corps is big on teamwork. But the thing I'd most like people to understand is just how much time and energy these drill instructors put in and how much they care about these recruits. There's a lot of tough love down here.

"Sergeant Major Arndt got an e-mail telling how eighty percent of one company was made of brand new Pfc.s and lance corporals and the writer said his NCOs just dove in, formed their squads, went over there, and they took only one casualty the whole war. That was in the beginning, of course, but he said those kids functioned exactly the way they were trained, and these are eighteen- and nineteen-year-old kids who three or four months earlier were down here whining about being homesick. So the system is working.

"We took more casualties in the initial stages of the war but we had the harder fighting. We went up the right side and hit the cities, where the Army had their end run around the left, over open ground, with their mechanized forces, while we slugged it out in El Nazirihya and Basra and other cities, and we still made it to Baghdad the same time the Army got there. That's what we were trained for: We started back in the mid-'90s with Gen. [Charles] Krulak with his Operation Dragon from the Sea, which was heavy on urban warfare.

"We've been practicing that around the country in Cleveland and L.A. and some of the rundown parts of some of the major cities. Helicopter assault on the rooftops, taking the buildings, street fighting—all that stuff. We have used some of the more built-up modern parts for practicing our reconnaisance; we've practiced sending Marines in civilian clothes on bicycles as tourists with missions to get pictures of traffic, of buildings, and the like.

"We started going toward a thing we call the Three-Block War. Where everything inside an urban environment is going to happen in a three-block radius, and we've been pushing authority down, like we always have, to the lowest rank, to corporals being fire team leaders and squad leaders and making these decisions in these confined areas.

"It was very successful in Fallujah, even though it's hard to tell an Iraqi insurgent from a citizen. Some of our contacts over there tell us it's not easy when someone comes out of a building firing a weapon and they're standing behind a child.

"It bothers me that I have been unable to get over there. It bothers me and it bothers my wife, Denise. When the war started, we went out to register our car and we gave the woman our registration and she said, 'Oh, you are two of the lucky ones who didn't have to go.' We said, 'No, we're two of the unfortunate ones who were left behind.' She said, 'Oh, I never heard it like that before.' I was irked. Why would you assume I would feel lucky that I didn't have to go? Same thing happened at a D.I. Association, when the wife of a retiree said, 'I bet you're glad you didn't have to go,' and I said, 'Every Marine wants to go over there; that's where the fight is; that's where our Marines are. It pisses us off that we're stuck here in a nondeploying unit.'

"I actually called the sergeant major up there, told him my wife and I wanted to go, and he said, 'You know how long the list of sergeants major that want to go over there is? We don't need you there. There are no sergeants major who are casualties: There's no one to replace.'

"I've always been interested in military history and getting a chance to go back to Okinawa and then getting to visit Iwo Jima made me want to learn everything I could about it. You get on the island and you see the control tower that says Iwo Jima, and you know you're there, but it really doesn't mean anything. You walk down the dirt road, and you see some monuments. The Japanese have these little obelisks with writing on them, and within fifty feet of those things is a cave opening or a tunnel, so you know there's something nearby but it doesn't strike you. You keep getting closer to Suribachi, and then get up on the road going back and forth, and you get up to the very top, and when you get to the top, you step to the edge and you look down on the beach and then it hits you. You'd be a liar if you said you didn't feel something. I stood there and I got a chill. The closest thing would be like a Muslim going to Mecca. It's a pilgrimage you have to make if you get the chance.

"Everybody knows the basics: twenty-six Medals of Honor, the Marine divisions that were there, the dates. But I wanted to know more. The fighting was so intense and just nonstop, thirty-six days, and the amount of casualties was just enormous. Six thousand eight hundred Americans killed in action, 27,000 total casualties, 100,000 Marines taking part. Twenty-one thousand Japanese dead. And when you get on the island and you look around and you think how could they have put that many people on this island without bumping into each other? You see the combat footage of it and then you go there. I am standing right where that footage was taken and people were lying here fighting and dying. There were pillboxes still in small bushes, hundreds of them. I stepped on one without knowing, it was so well camouflaged. And the sand, the sand is volcanic ash, you stand on it and you sink. You try to run in it from the beach and in six or seven paces you're breathing hard. You dig in it and it just falls back on itself. It's not like regular sand, which is square or cubed. It's an odd shape and it just slides.

"I give talks on base to various units or to Marines who just want to know. It's motivation, and it's history. It takes about three hours to get into the battle itself and tell where the units went and what happened. It's the only way to get full comprehension of the enormity of the battle itself.

"The tunnel and cave networks went for miles underground. Poison gas was considered but then was ruled out for fear of retaliation. The Marine Corps asked for ten days of naval gunfire and the Navy gave us three, saying that would be enough. But I don't think even ten days would have made much difference. They went underground and stayed there. I've been in some of those tunnels. They are hundreds of yards below the surface. You can still see hundreds of thousands of pick marks in the tunnels left by the Korean laborers who were used to carve them out. The ceilings are arched and most Americans have to crook over to get through them. Originally they mined sulphur there, for matches and things. The name Iwo Jima means Sulphur Island.

"It wasn't until the summer of '44 that the Japanese realized that Iwo was right in our path, and they didn't take out the last civilians who lived on the island until December. There were three villages that grew sugar cane and vegetables, and the people in them were evacuated to a northern island called Chichi Jima. They only had eight or nine months to get ready for the invasion. The island is shaped like a pork chop, only five miles square, with Suribachi in the bottom corner of the skinny part. We attacked from the southeast, on two sets of beaches. It's honeycombed all though and they still discover new tunnels every year. A Japanese organization goes to the island and searches for new tunnel openings and they document all the new tunnels they find. They look for artifacts or bones.

"The island was returned to Japan by the Johnson Administration on June 26, 1968, in very low-key fashion. A lot of veterans I talk to now still think we should have kept the island. They don't like the idea that it went back because of what it is, what it represents. Iwo Jima belongs to mainland Japan—the prefecture of Tokyo. It would be like them capturing the Hawaiian Islands and never giving them back. It's 700 miles from Japan, just an airstrip. They don't even have fresh water. They make their own."

CHAPTER TWENTY-FIVE

JOSH WYLIE

Sergeant Major
Recruit Training Regiment
Parris Island
June 1977–

I don't think I've ever really had any issues with
discrimination throughout the Marine Corps. I guess it
goes back to my upbringing. . . . I would have to say, if I have been
discriminated upon throughout my life, I wasn't aware of it.

Sgt. Major Josh Wylie, Recruit Training Regiment,
Parris Island. He told his wife, Wanda, prior to their
marriage in May of 1984 that they would have to wait
because he was "dedicated to my recruits, making
them Marines, and I wanted to finish with them
first." (*Photograph by Cpl. Brian Kester, Combat Correspondent, MCRD Parris Island*)

273

I met Sergeant Major Wylie a couple of times during visits to Parris Island, and we ended up with a long conversation over the telephone, which was the basis for this interview. I was fascinated to hear him declare he had never encountered racial discrimination during his time in the Marines, which makes the contrast between today's Marines and the Montford Point era of the 1940s all the more remarkable.

"I could have went to college but I just didn't have the desire and I didn't want to jeopardize wasting my mother's money with me not putting out, so I decided to go in the military. I grew up in Rock Hill, South Carolina, three hours from Parris Island, but I had no exposure to the Marine Corps, so I really didn't know anything about it, other than this friend who lived behind me. He had joined the Corps. I was in an Air Force ROTC program in high school, but just seeing him in the uniform and hearing everybody talk about how hard the Marine Corps was, that decided me. I figured if I was going to go in, I wanted to go in the toughest one. And everybody said that was the Marines. And, besides, they had the best-looking uniform.

"I'm six feet one and right now I weigh 195, but I wasn't that big growing up, in high school, although I had the ability to fight off blocks and get in and make the tackle. I played defensive end for the Rock Hill Bearcats. I was voted most valuable player one year. We had winning seasons but we didn't make it to the playoffs.

"I signed up the previous December and went delayed entry in June of 1977, right after I graduated from high school. It was scorching at Parris Island. Physically it wasn't that hard for me but it was still challenging. I dealt with the sand fleas just like everybody else: You stand there and let 'em eat. Boot camp was difficult, but it wasn't anything I couldn't do. It was pretty much like it is today. One difference is they have a fence around the Confidence Course now for the alligators, to keep them out of the pool of water where you practice on the Slide for

Life. It's a rope you grab onto where you swing your feet up and shimmy down feet first. Actually, you start out lying on top of the rope with your feet out. They teach you a way to curl your foot and one leg hangs while the other foot curls around it.

"At a certain point you are required to reverse and crawl under the rope with your feet resting on it and using your hands to pull you down. When you have to reverse there is always a tendency to fall off the rope into the water below. I guess a group went to do the course one time some years ago and there was an alligator in it. When we heard they put that fence up, we laughed because it was just another part of the thing drill instructors would give recruits a hard time about.

"Of course we didn't have the Crucible then, but you always went out to the field at a certain point in boot camp and bivouacked and you went through the infiltration course, and the other things. When I got to be a drill instructor out there, we'd place quarter-pound sticks of dynamite in the pits and we'd blow the pits to provide live explosions when the recruits crawled through. Now we use simulators and all that, obviously for safety reasons. It wasn't called the Crucible but we did a lot of similar things.

"General Krulak pulled it all together, added some mentoring parts, and it's very challenging for recruits these days. I do honor grad lunches every Wednesday with the recruits, and I'll tell you it is special to hear these young people talk about recruit training. Often their high point is the Crucible because it is extremely challenging and also something they do as a team. They have to work together to get through and it causes them to bond.

"I graduated from recruit training as a private first class on September 27, 1977, and went to Camp Lejeune, to the Second Battalion, Eighth Marines. They stamped my book OJT, for On the Job Training, so I was going to be an infantryman. My unit had to spin me up, teach me what I had to learn about being an infantryman, and to designate my primary MOS. After six months of training in various tactics and weaponry, they made me an 0351, which is anti-tank assault man in a weapons platoon.

"We went on several Mediterranean cruises, what they called goodwill floats, and put on demonstrations in port after port, performing

helicopter landings and doing extractions. Every three days we'd pull
into a port and do a demonstration. We called them dog and pony
shows. We'd open up certain parts of the ship, have weapons dis-
played, and people would come aboard and we'd explain the different
weapons systems. We were like ambassadors for the United States. It
was a lot of fun. We got tons of liberty, went to a lot of spots in Africa,
Spain, France, and then we went to Brazil. On one particular float I
think President Carter had to go to Nigeria and we wound up going on
station to provide security for that.

"I got promoted to lance corporal within a year—it might have been
thirteen months—of being in the Marine Corps. After reaching corporal and then being promoted to sergeant, I requested orders to the
drill field, and came back here in January of 1982. D.I. school was
eight weeks, a very interesting place. On the one hand, it was like
going through boot camp again, except you knew you weren't a
recruit. You had liberty. But it was an eye-opening experience that
probably changed the whole way I looked at myself as a professional.
It was the little things. Anybody who gets orders to drill instructor
school is probably within the top 10 percent of their MOS, so they're
pretty squared-away Marines. I know I felt that way about myself. I
had been meritoriously promoted several times, knew I was doing
very well, but then, at the school, they just ripped you apart about the
slightest little thing on your uniform. My awareness of the small
things became very more apparent to me than they ever had been. It
just taught me you had to pay particular attention to everything.

"You had to learn the Standard Operating Procedure, you had to
learn the drill movements so you could explain them, not just demonstrate but talk through it as you demonstrated. You're given several
drill movements to learn, but you don't know which one you're actually going to have to stand up and do. And then there's still the physical thing, and, again, because I had been an athlete, it wasn't that big
a deal for me. But it wore you out.

"I still had not decided whether I was going to be a career Marine,
even though I had five years in by the time I went to drill instructor
school. When I went and signed up, I knew there were drill instructors
who got into trouble at boot camp and got relieved. So me, as a young

Marine, I'm trying to prepare myself, because you never know. You don't go down there with the intention of getting in trouble, but it is a place with a lot of rules. In the back of my mind was the concern that things could go wrong. I had only two years left on my contract at that point, and my basic attitude was, I'll do those two years, and, if I survive, I'll go on from there. I ended up getting meritoriously promoted to staff sergeant after a year. I became a senior drill instructor as a sergeant, and that was pretty rare.

"I had a year left on the drill field when I met my wife, Wanda. My mom and I were very close, and I had come home for a visit. She was very big in the Elks organization, and they were having a dance, and she asked if I wanted to go. I told her I would go with her, but I really didn't have anybody to go with. So my niece set me up on a blind date with the person which is now my wife, Wanda. We waited until May of '84 to get married because I told her—it was kind of selfish of me, to a degree—but I told her I was dedicated to my recruits, making them Marines, and wanted to finish with them first. I detached from Paris Island here in April and we got married in Rock Hill on May 19.

"She went with me after I left the drill field for the Marine Barracks PLP [Personal Liability Program] in Guam. It's a program where you guard special weapons, nukes and so on. I was a guard chief slash platoon sergeant. We have a base in Guam, but it's used now more as training, but at that time it was a big place where we stored and housed nuclear weapons. They were the big talk of the town then. All that cold war stuff. Wanda and I have been married twenty-one years. We have a daughter, Deonna, who will be twenty next month, and our son, Joshua, will be eighteen in a week. Deonna is in college right now. We're extremely proud of both of them.

"After my tour on the drill field worked out, I ended up having to reenlist for three years, which was going to put me at ten, so I decided I might as well stay for twenty. The thing that attracted me most to the Marine Corps was that you were secure with the job as long as you performed. Equally important was the friendship, the brotherhood you experienced as a Marine. I have some friendships where I may not hear from the guy for years or vice versa but immediately, if we cross paths, it's like there has been no time in between, like I saw him yesterday.

"As for discrimination, for me personally, I don't think I've ever really had any issues with discrimination throughout the Marine Corps. I guess it goes back to my upbringing. My first three years as a kid I went to an all-black school, and then the big integration movement started and we ended up getting bused to schools where white students went, from fourth grade on. My mother wasn't a person that was hostile or racist in any manner. She preached education and getting out there and taking care of yourself. She preached tolerance, understanding, and self-control. And she had some hard jobs. She also had one job where she worked for this white family and they treated her very well.

"My mother's name was Sarah. She died at the age of sixty-seven in 1988. It took me a long time to get over her death. She lives in my heart.

"But, anyway, it was almost kind of normal for me to be with different races, so it didn't bother me at all. Three of my sisters became schoolteachers. I would have to say, if I have been discriminated upon throughout my life, I wasn't aware of it.

"Earlier in my career, as I was being promoted, I often had young Marines coming to me saying they weren't getting a fair chance, a shot at, let's say, meritorious promotions, because they were black. They would come to me because I was a black person who had done well. The first thing I would do was look at their record books. And I'd question them on things such as their PFT [Physical Fitness Test] score. Were they a first-class PFTer, or third class? I'd say, before you blame someone for holding you back, you got to make sure you've done everything you could to qualify. And if you haven't, then the problem is you. It's not the other people you are trying to point the finger at. If you have done all the things you are supposed to, if you got top scores in everything and your performance is outstanding, and you are not being recommended for meritorious promotion, then there may be a problem. In most cases, I found it was a matter of educating them to understand: You can't think that somebody's going to give you something. You got to work for it. I wanted to instill that in them.

"I can't say the military is fairer than the corporate world when it comes to advancement, because I've spent all my adult life in the military, but I do think the military seems more open. I found the Marine Corps was about people performing, doing their job. And they didn't care what color you were, or how tall. As long as you met the standard and you performed, you're all right. If you're a nonperformer, you ain't going to get promoted and you're going to get weeded out. The military is not a place where they measure by skin color.

"I liked different challenges and I constantly looked for things that made you stand out. In 1989 I put in to be an instructor for college students in naval ROTC programs. I applied to become an assistant marine officer instructor, got accepted, and was sent to Hampton Roads, Virginia. My primary office was at Hampton University, but I also worked at Old Dominion and Norfolk State, teaching college students who had signed up for Reserve officer training.

"In the summer, when the college was closed, I would go serve as a drill instructor for Officer Candidates School, teach drill, the same things I did as a drill instructor at Parris Island. From there I went on several deployments with the Third Battalion, Sixth Marines, to Somalia, right after the Blackhawk Down incident, and to Haiti and Guantanamo Bay. Later I did instructor inspector duty with Otis Air National Guard on Cape Cod, and in November of 1998 I was promoted and transferred to a Harrier squadron in Yuma, Arizona, as the squadron sergeant major. I was there three years when I got orders back to Parris Island as the sergeant major of the Weapons and Field Training Battalion. After less than two years they wanted me to come and take over the regiment, which is comprised of the three men's recruit battalions and the women's battalion. I came back here in 2002, almost three years ago. This month, on May 27, 2005, I will transfer to TBS, the basic school, at Quantico as sergeant major for the training of new officers.

"As the regimental sergeant major, you are extremely busy because you are dealing not only with recruit training but also active-duty Marines, your drill instructors. Standards are so very high there are always going to be violations in some fashion to the SOP, the Standard Operating Procedure, and we look at those individual violations and we deal with them in whatever way we feel is necessary, based on what occurred. Right now the jury's still out on the drowning of the recruit Jason Tharp on February 8, because the investigation hasn't been concluded, although several Marines have been suspended in the meantime.

"I missed Desert Storm because I was on ROTC duty in Virginia. They came out and told us, Don't ask, because you're not going. It was the same with Iraq. When 9/11 happened, I was out aboard ship, but we didn't get near that area. Every Marine serves a purpose: You are either in combat or you're doing things that are supporting. We have talked to a lot of the drill instructors because obviously they wanted to go to Iraq. We had to actually explain to them that their jobs here have become more important because they have to train these recruits. It can be frustrating, but you just have to make people understand they are contributing when they may not feel they are doing so.

"The biggest thing I try to do is to get the drill instructors to understand they live in a fish bowl and that people notice what they do. And something bad gets really noticed. That's my big focus down here, really, is to try to make them understand that fact that they do make a difference. My battalion sergeants major and the first sergeants who work for them are my information source. I meet and talk with them daily. We also have a regular weekly meeting unless I need them prior to that. I talk to each of them daily by phone or in person. They come by my office and I go to their offices. I have a regular weekly meeting with all my sergeants major and we talk individually as things come up. I also try to meet once a month with them and all the first sergeants. I think the system works. I could not possibly do this job without those sergeants major out there in the battalions. They know they can be open and frank with me about whatever it is they want to talk about. They understand that I might not agree with what they say, but they know I want them to be open and frank. I want to know what their thoughts are so I can help them make the best decisions.

"As the sergeant major of the Recruit Training Battalion, I oversee five battalions, each of which has its own sergeant major. Besides the four training battalions, there is the Support Battalion, which has all of our classroom, swim, and martial arts instructors. Support also has the Receiving unit, which handles incoming recruits, the medical rehab platoon, the PCP, and EHP, a holding platoon where we are deciding what to do with a recruit, whether to send him or her home, or back to training.

"The Support Battalion has no permanent drill instructors assigned. They go there after having served at least a year in the training battalions.

"There are a total of nine sergeants major aboard the base. Me and my five, along with the Weapons Battalion and H&S sergeants major, all report to the Depot sergeant major, who reports to the commanding general.

"Obviously, times have changed in the twenty-eight years since I went through as a recruit, and the changes we've made have been for the better. We are just adapting to the different ways we are fighting battles today, and there's a lot of information being given to these recruits. But, if you really look at it, you see that recruit training still remains the same.

"The thing I'd most like people to understand is just how much time and energy these drill instructors put in and how much they care about these recruits. There's a lot of tough love down here."

CHAPTER TWENTY-SIX

ROB BUSH

Sergeant Major
Third Battalion
Recruit Training Regiment
1983–

I believe that Parris Island is really about belief.
And I think that kid who barely makes it across, who makes it
through the wire there, the kid who really had to fight to be
here, could turn out to be a superstar in the Fleet."

Sgt. Major Rob Bush, Third Recruit Training Battalion. Dave
Robles (Chapter 5) was his battalion sergeant major when
Bush was a recruit. Bush said he looked at Robles and said to
himself, "I want to be that guy." He said Robles "didn't know
me, I never met him." Years later, they talked. Robles, the
sergeant major said, is "a pretty good guy. He's good to go."
(*Photograph by Cpl. Brian Kester, Combat Correspondent,
MCRD Parris Island*)

I met briefly with Sgt. Major Bush during the Drill Instructor's Reunion at Parris Island in the spring of 2005, and then we talked later over the phone. Considering his length of service and experience in the Marine Corps, I wanted to know how he perceived the mystique of boot camp and what it was that turned recruits' lives around in such a relatively short time. He said the important thing was to induce recruits "to buy in" to the Corps and its beliefs. Another of his jobs, he said, was to remove "the nostalgia myth" and discourage drill instructors from thinking that today's Marines and recruit training are not like "the good old days." His career-long identification with Dave Robles reflects the continuity that sustains the Marine Corps.

"I was born in Colquitt, Georgia, on January 29, 1963, and graduated from Riverview Academy in Albany. I may be the only Marine that ever came out of Riverview. My mother was a civil service employee on the Marine base in Albany, but I actually never went there. I did a year in college, and the whole time I was there—Ronald Reagan had become president—things that were happening in Lebanon and those places, the transition between Carter and Reagan and the Iran hostages kind of put my eyes on what was going on in the world. I had heard about the Marines.

"The summer after my freshman year my parents were asking me what was wrong because I didn't look like I was too excited about school, and I told them I really wanted to be a Marine. They asked me why I was going to college and I said, I thought you wanted me to go, and they said, No, we never told you that. So I said, Well, I'm going to see a recruiter tomorrow. That was in August of '82. I shipped down here to Parris Island, delayed entry, in January of '83.

"I didn't have any real grasp of the difference between an officer and an enlisted man, just a rudimentary understanding that Parris Island was the place where they made Marines, and that was where I wanted to go. I had played football and ran track in high school, the two-mile. Boot camp was good except I got sick in recruit training, bronchitis and pneumonia, but I was the Guide, the guy that carried the platoon flag up front, so I didn't feel like I had the option to back down. I was never taken out of training. I just did a lot of throwing up after meals and sucking it up. We didn't have the Crucible. I wound up the honor grad of my platoon.

"After graduating from boot camp I went to electronics school with the air wing at Millington, Tennessee, for eight weeks, then got sent to Lowrey Air Force Base in Denver for ten months of study with precision measuring equipment, all the equipment they used to work on airplanes. We calibrated and repaired that equipment.

"Then I shipped to Okinawa and went to NCO school there at Camp Hansen, got meritoriously promoted to corporal, and then after a year there, I came back to Cherry Point doing the calibration job and got picked up for meritorious sergeant. I reenlisted a year early and for my duty station option I chose Parris Island, and came here as a drill instructor in 1986.

"I was told over and over that my job, my MOS, was technically critical and they didn't like letting people go, but my gunny thought I'd be a good candidate for drill school, so I took my record book down to the drill instructor screening team. The Parris Island Recruit Depot sergeant major, his name was Johnson, was with that team visiting Cherry Point, and I happened to be in his line. He looked at my book and said, Sergeant Bush, why do you want to go to Parris Island?

"And I told him that, from the first time I saw one, I always wanted to be a drill instructor. In fact, that guy's still here, although he's retired now. His name is Robles (see Chapter 5). He was my battalion sergeant major when I was a recruit. While my battalion commander was giving me the Band of Brothers speech, Sergeant Major Robles was standing off on the side, and I never heard him say a word, ever, but I looked at him, and this is an absolute true story, and I said to myself, *I want to be that guy*. I want to be that guy right there, that sergeant major right there. And I saw Robles in 2003 when I got back down here as a sergeant major—he didn't know me, never knew me, I never met him. He came to the Marine Corps ball and I went up and introduced myself to him and his wife and told him what I just told you and we sat there shaking hands and I told him, I guess I made it. He's a pretty good guy. He's good to go.

"I went to D.I. school in March 1986. The school was very arduous, but I had a good squad adviser and I thought the curriculum was good, I thought the drill was good. It was a natural fit for me. I had the highest average on record when I graduated, 99.3 or something.

"I think we started with 103 and we graduated 51. I wouldn't characterize myself at the time as happy about that, but I was glad to see that the process narrowed us down to guys who came there with the fundamentals all ready, and they just reshaped them for us. I was a little bit 'belt fed.' My expectations today are a little different from what they were then. I was a lot more rigid and I saw individuals who really had to struggle. I thought they lacked the prerequisites. It was a struggle for me to think that some of them were going to be drill instructors and in my young mind at the time I wasn't sure they were qualified. Dealing with that in the squad bay was probably more of a challenge for me than anything else. I didn't challenge their motivation. I was troubled by their makeup. For example, the Marine Corps doesn't require you to max out the PFT, the Physical Fitness Test, to be a drill instructor. I maxed out the PFT. Other people were running in the 290s, very high, but if you could do above the minimum, you were going to make it through drill instructor school. As a young sergeant I looked at that and thought we ought to be giving these recruits guys who maxed out, drill instructors they could look up to and say, like I did with Robles, I want to be like that guy. You didn't want to have recruits showing up trained by their football or wrestling coach and in better shape than the drill instructor.

"That bugged me a little bit. I also thought a Marine ought to look a certain way, bigger in the chest, good arms, good legs, good shoulders, handsome, all that crap. That obviously was a skewed paradigm, pretty judgmental and unforgiving. In retrospect I was wrong. My motive wasn't wrong. My motive was pure. But my paradigm was a little bit skewed.

"I was on the drill field two years, from 1986 to 1988. Each team was generally three Hats; sometimes we'd have four. There was always emphasis on no physical contact with recruits. That was always the case, and I don't think what's illegal today is different from what was illegal then. It has never been legal to dump on or abuse recruits. You got guys that do thump on someone and in my characterization that guy's just a criminal, and he ought to be court-martialed.

"But you also have drill instructors who really care about what they're doing, and they really want this recruit to get it. They can overstep themselves when they do not get the desired result. The intent is never malice, never to hurt that kid or to move him aggressively, but it happens. And you can walk your way from there all the way down to the drill instructor who would not think of touching a recruit. And really where a lot of the noise comes from is when drill instructors have been seen to touch recruits, adjust their gear, adjust their uniform or whatever, and then in somebody's eyes it is perceived as more of a threat than it really is. I also think that by the time the recruit gets out of training and goes through an enlistment or two, the stories are embellished and take on a life of their own.

"It is very demanding to take a young man from Anytown, U.S.A., put him in one of these squad bays, and push him through our process for twelve weeks. It is very demanding on the drill instrucutor. It's an eighteen-hour day, a brain-, body-, soul-draining siege. I did four cycles as a Green Belt, a junior drill instructor, and then I was a sergeant senior drill instructor for two and a half cycles, picked up meritorious staff sergeant halfway through my seventh cycle, and then I did one more cycle as a staff sergeant. A cycle is a twelve-week rotation, from pickup to graduation. And, yes, recruits still arrive on the bus after dark.

"I went from Parris Island to the air station in Beaufort, working in my MOS again, and then Desert Shield/Desert Storm happened, and I deployed to the Persian Gulf and when I returned in a year, year and a half, I went and taught electronics, calibration, at school in Albany, Georgia. I got promoted to gunnery sergeant and spent a couple years there as chief instructor, then got selected to teach at the staff NCO advanced course at Camp Lejeune. I was instructor of the year there, from '97 to '98.

"Then I got selected for first sergeant and sent to the Second Battalion, Sixth Marines, an infantry battalion at Camp Lejeune, and I was with Echo Company, Sixth Marines, from July 1998 to March 2001. I did a Mediterranean deployment with Two-Six, as first sergeant on the USS *Oak Hill*, one of three ships on the cruise. We did a lot of exercises, a regular six-month deployment. We went into Kosovo for a little less than a month, just south of the Serb border. We went to Spain, Italy, France, Greece, Israel, Croatia. We got liberty in all those places. Liberty's good.

"I came back from that deployment in March of 2001 and went to Headquarters and Service Company, same battalion, Two-Six, and did a deployment with them down into the Persian Gulf because 9/11 occurred. We went into the Suez Canal, Red Sea, Gulf of Aden, and then we were doing mostly Gator Squares off North Africa, just going in circles, off the coast of Djibouti.

"After we returned from there I became the acting sergeant major of Two-Six in December 2002 at Camp Lejeune. I was selected for sergeant major the following month and then deployed with Two-Six to Iraq near the beginning of February. Two-Six came in as the last infantry battalion in the country. I had a company that was a security element for the MEF, the Marine Expeditionary Force. General Conway was the commanding general and his security detachment was one of my platoons, combined with armor teams, Humvees with 50 cals [50-caliber machineguns]. We were always a little bit behind where the MEF main forward elements were. Our guys did POW missions, urban patrolling. We went from there to Okinawa. Then I got orders to Parris Island in July 2003, as the Third Battalion sergeant major. I report to my commanding officer and I also have relationships with the regimental sergeant major [Josh Wylie, see Chapter 25] and the depot sergeant major, helping shape or advise what is going on at the battalion level so they can make better policy from their levels.

"In many ways, the noncoms, the noncommissioned officers, run the Marine Corps. The gunny is the man. I think that awareness kept me from going to Officer Candidates School. I got asked about going to OCS all the way until I was a gunnery sergeant. As a young first sergeant I had officers even when I was at P.I. trying to get me to put in packages for it, and, it sounds a little hokey, but it's the truth: I said when I was sitting on that Parade Deck and Sergant Major Robles was standing there, I said that was who I wanted to be. I just never wanted to be a commissioned officer. I don't have any regrets about that.

"The young Marines in Iraq performed superbly. I have not done any extensive reading about recruit training in the World War II era, but it is my understanding there were times during that era when recruit training was short and tight, eight weeks, and a lot of guys learned how to use their weapon on the ship on the way over. This is not to say recruit training is harder today. The difficulty is not with or without a Crucible or twelve weeks or eight weeks, whether you have new Pfc.s as drill instructors, as they did in the past, or if your drill instructors have been NCOs in the Fleet and have come back.

"I believe that Parris Island is really about belief. And I think that kid who barely makes it across, who makes it through the wire there, the kid who really had to fight to be here, could turn out to be a superstar in the Fleet. It's like getting a guy on the Yellow Footprints and when you look at him, he's got this roll of fat hanging over his belt and he's kind of sloppy looking and he just hasn't done anything his whole life. But you know the reason that thing's hanging over his belt is he had to lose fifty pounds just to get here, and to me a young man who's willing to put his hand in there when there's a war going on, who had to really bust his hump just to get on those Yellow Footprints, is worth the best we can give him. And if he fights his way through the twelve weeks of recruit training and barely makes it across the wire, there's one thing that guy's already demonstrated: And that's that he believes. And that he's got the heart.

"That may sound a little hokey too, but I say it to my Marines here all the time in battalion, that what we do here at the Emblem Ceremony on Training Day 69, our graduation on Training Day 70, is turn out Believers. They're just baby Marines. We can call it basically trained, we can talk about all the stuff they've been through, but just like anything else in life they may forget 90 percent of that stuff. They're not going to forget that they're a Marine. They don't forget that they're a Marine. They don't forget that kind of indelible kind of unable-to-define-what-it-is thing it is to be Marine. And they're going to go to places like Iraq, and they're fighting for their brothers. They're not engaged in Fox News or CNN or what's the latest poll numbers for whether we ought to be there or not. They're fighting for each other. The man on the right and the man on the left. It's always been that way. It's almost become a cliché but it's not. They believe in each other. I think that's what we do here.

"I think it's more akin to magic or voodoo; if you could just take the training schedule and say, it's the training schedule, then you could give it to anyone and they could all become Marines. But that's not the case.

"And when you say Platoon Whoever they are, Dismissed! you could follow that up with a, 'And so it begins.' I mean, I believe it begins at Parris Island for sure. I think that what these drill instructors do here, what they've always done here, is very difficult to define. Very difficult, I think, for them to do. I see them all the time and, even now, I look at them and it feels like, I've told my Marines this, I told my first sergeant this for sure, that every day they walk among giants here. I think these guys, whether they are the best where they came from or whether they had to fight just to make the grade, you're getting everything they got. They're giving you all they got. And some guys got more than others. You can't ask more than that from a man.

"They buy in. I think that's a good word. That is exactly the transformation. They use that word a lot these days. And I have to say you have guys who never talk about anything but the Marine Corps. Here, my friend, take a breath [meaning, don't overload being gung ho over the Marine Corps].

"As for the Crucible, my professional opinion is, if that's what the Marine Corps wants, to implement another tool down here at Parris Island to try to institute a higher degree of belief, then I'm for it. I think it's a good thing. I think we got recruits who go through it and I think that's part of the belief-maker. But I don't think it's a deal-breaker in that, if we didn't have it, we wouldn't have what we got down here. It has been played up as a rite of passage, and it obviously doesn't hurt, but I don't think we would lose anything in its absence.

"The magic of Parris Island ain't at the Crucible. It's happening in the squad bays and on the Parade Deck and on the PT field. And what you throw at them for classes. We have a lot of classes today that maybe weren't taught when I was here, core value classes or classes on sexual harassment or operational risk management. I don't think those are bad classes. I don't think it's wrong to expose a recruit to those things. I'm not sure it's actually retained at this stage in the game, but it's another facet to what it is to be a Marine: This is what we believe. I really think it's more art than a science. I think if it was a science we could take this training schedule and say, Here, do this, but there's far more to it than that. There are nuances very difficult to articulate.

"As for the women in the Marine Corps, I would have to characterize them as falling into extremes. They're either extremely proficient, by which I mean that they excel to a degree seldom matched by a substantial portion of male Marines, or you have women who do not perform at that level. I think it's possible they are more visible just because there are fewer of them.

"The superstar females tend to excel to a degree that makes them competitive with the men in many areas. I mentioned I came out number one in my D.I. school class; the number two graduate was a woman. I've seen the women drill instructors from a distance here and it seems to me that they are demanding almost to a degree that's over the top. I don't know if that's a conscious or a subconscious thing, whether they're competing for acceptance with the men.

"The women Marines in Iraq are doing just fine. And that's not just the party line, either. I've seen over and over again in difficult circumstances young Marines, both men and women, doing amazing things, sometimes when they're not even well led.

"As sergeant major I have four companies with six platoons per company. Each platoon has three to five Hats. On any given day my morning report is different, but that's how it is structured. So I have from 90 to 115 drill instructors, and that's not the total number on deck, because every sergeant major sends drill instructors out to run the rappelling towers, take care of close combat, work the swimming pool and the academic instruction.

"The guys that run Receiving, that do the Yellow Footprints and do the processing on the back side if recruits get dropped, are all drill instructors on quota. I talk to my drill instructors regularly, and maybe

I'll bring up some of the things that they are doing, a trend or maybe an isolated incident, some violation of the SOP, for example. I spend a lot of time talking to them about why are you doing the things you're doing, why do you think what you think? What causes you to believe what you believe as a drill instructor? Where you'd get it from? Did you get it from a movie, did you get it from the guy that trained you? Many times they'll come out of drill instructor school with a preconception of what a D.I. is supposed to act like or be like, and sometimes I find them to be way off base.

"I'll ask them, Why do you think that? What I want is for them to be good leaders, good role models. I want them to be very exacting and accept nothing less than better. Whatever the recruit is giving, get more from him. But that doesn't require mindless yelling or mindless repetition. It needs to be very mindful application to change that kid's mind to make him more like us, to make him a Believer by Training Day 70.

"Sometimes you'll see D.I.s on the Parade Deck or in the squad bay running up and down the line screaming and yelling to move faster, faster, faster. They might do that half a dozen times. And the recruits are not technically moving any faster. So the drill instructor's getting stressed out. The recruits have figured out how fast they can move and it will still be acceptable so some drill instructors believe if they move faster or sound louder, they'll get more out of the recruit. I disagree with that philosophy. I think you got to watch what the recruit's doing, and let the recruit know he's not doing what I want here. Sometimes that's with volume, sometimes that's with speed, but the bottom line is the recruit has got to be kept out of his comfort zone. But that doesn't necessarily mean a lot of running up and down yelling and screaming.

"I have watched drill instructors correct a drill movement a dozen times because four recruits were in error. So, the old thinking is, if four of them mess up, we all do it again. Well, if you're trying to get better performance out of them, which is the goal, I think, instead of punishing the whole platoon by making them redo the drill movement twenty times, you pull the four out that are messing up and let the others go on because the truth is about 80 percent of them got it right; 20

percent ain't getting it right. If you make them do it ten times the ones that were getting it right don't give a damn any more. So you dumb them down trying to get the others to come up. We talk about that kind of thing.

"I've got drill instructors here who have set a new regimental record by doing what I'm saying, so I think the application is valid. A recruit who is out there shaking like a leaf, petrified of the drill instructor, is not going to give you his best performance. He's not listening. He's afraid. I tell my drill instructors we're here to bend these guys, not break them. Everybody bends at a different spot at a different time, at a different place. And when they bend, your goal is let them get strong right there, and then bend them some more.

"I don't believe recruits today are any less than they've ever been. If I were to compare what I see today and what I saw on the Footprints when I got here, there's no significant difference. Drill instructors like to think, Boy, it's not like it was when I was . . . but none of my active drill instructors were even in the Marine Corps when I was a drill instructor, so I can look at them and say, Don't tell me what the hell it was like when you were a recruit, because I was a drill instructor before you were a recruit.

"One of the challenges for the drill instructor is to keep in mind his ultimate goal and figure out how to put out the best Marines he can. But he can't lose sight of the fact that someone who meets the minimum Headquarters Marine Corps standards on graduation day is qualified to be a Marine. It's hard to find a young sergeant just out of drill school who buys that. He's like Sergeant Bush coming out, saying, Hey, guys, gotta go. But that minimum standard recruit may turn out to be a superstar one day. The True Believers coming out of drill instructor school can't lose sight of that, and my job is in support, to help them keep that in mind. I say that all the time, to the other sergeants major, We ain't the job here. The guys wearing the hats and the belts are the ones with the job.

"It's all about buying into the system, but if I lead as a sergeant major or if my drill instructors try to train to some unattainable mythological level that never really did exist because they have taken reality and exaggerated it, well, you're going to do a lot of crazy things, you're going to wear your drill instructors out, and you're going to hurt some recruits.

"Jason Tharp [the recruit who drowned during swim training at the base pool on February 8, 2005] was in this battalion, and I was pretty attached to that whole incident." (The investigation was incomplete as of this writing.) "It was obviously a tragedy no matter what comes out in this investigation. I think about his parents, and I recognize that word is pretty weak. Nobody down here wants to hurt one of these kids. No one wants one of these kids hurt while he's down here.

"I think we're doing a good job continuing on with what we normally do while at the same time trying to be sensitive to the fact that we have had this tragedy. And while incorporating all that, trying to relook at what we're doing, I don't think there's anything wrong with trying to second guess yourself. I think it's healthy. I think it's healthy to look back and go, What should we have done, if anything, different? I think to go into a siege mode and say, We didn't do anything wrong, is not the right approach. And at the same time it's not necessarily true that something must have gone wrong. When this investigation comes out, we'll see what it was. I say this pretty regularly down here, also: Truth wins. Whatever it is, we'll deal with it.

"Things go to hell in every area, in my opinion, when you try to cover up. Even if you survive the damage control part of it, even if you're successful with some kind of damage control, I think you still lose, because now you can't change what needed to be changed. Because you got to buy your own damn lie now. The truth coming out allows us to fix something that needs fixing, if that's the case. And it will allow us to be comfortable that we don't have anything broke, if that's what comes out of it. Hell, I second guess myself all the time.

"My term here expires July 2006, but I could leave anytime, for wherever Headquarters Marine Corps wants to send me. Once I get twenty-four months on station here, I'm movable. I could put retirement papers in now, if I wanted. Thirty years is the maximum, which gives me about eight more years. Some days I'm thinking along those lines. One of the things I see in my former peer group, sergeants major who are now retired, I don't think there's a stereotype, but sometimes it is very difficult to be detached from what you've always been. I'd like to teach maybe at the college level, I'm not sure. Something I'd like to do. I've got about two semesters left before I'll have a bachelor's degree. I've been doing it all along. I intend to be that guy who vigorously pursues another career.

"I think I'll always be a Marine. I have no desire to be separated from being a Marine. It was a very emotional thing for me, when I was a senior drill instructor and I said, 'Platoon 2000 whatever they were, Dismissed!' It was a big deal. I can sit up there in the reviewing stand every time and when I'm watching them put that Eagle, Globe and Anchor in that kid's hands, and he's putting it on his cover, and he puts it back on his head, in my head I'm making that statement: 'And so it begins, my friend.'"

CHAPTER TWENTY-SEVEN

WILL POST

Sergeant Major
Second Battalion
Recruit Training Regiment
1981–

Another thing I always preached was about killing somebody.
I always believed the bottom line is, the Marine Corps exists to
kill people. You can filter through all the political rhetoric
but the bottom line is, we exist to kill the bad guys.

Sgt. Major William Post, Second Recruit Training Battalion. He led Marines in combat in Kosovo and Afghanistan. "You always hear about how much tougher the old guys were in the old days, but it is NOT true," he declared. "When we went to Afghanistan, these Marines were carrying . . . at least seventy-five pounds per man." (*Photograph by Cpl. Brian Kester, Combat Correspondent, MCRD Parris Island*)

I met both Will Post and Rob Bush at Traditions on Parris Island at the welcoming breakfast of the annual reunion for former drill instructors. Active drill instructors, series officers, and sergeants major appeared to go out of their way to greet and befriend the old-timers, treating them with great respect and turning out in good numbers for their cadence-counting and story-telling contests. Marines, regardless of rank, invariably address civilians as "sir."

"I came in as a young buck the week I turned seventeen. I'll be forty-one on June 15 and I'll go over twenty-four years in the Marine Corps on June 22, 2005. They're going to do four weeks' physical therapy on my back, see if that does anything. If not, they're going to talk surgery. I've got to get that fixed because that's pretty much a career-ender there as far as being a Marine goes. Without your feet or your back, you can't do much, especially if you're a sergeant major. I got here April 1, 2004. I've been the battalion sergeant major one year, one month, and ten days.

"I'm from Milwaukee. I was a high school dropout. It was actually the only thing I've ever quit in my life. It was probably a good idea, because the Marine Corps really straightened me out and got me on the right track. I eventually got a GED, but I bet a good 25 percent of the Corps was high school dropouts when I came in. At the time, I was working and I couldn't sit still, couldn't focus. I grew up without a dad. They were divorced. My uncle had been a Marine and he was the father figure in my life. He was a Korea veteran and he always talked about the Marines. I saw a Marine recruiting billboard on my way home from work one night and I just stopped in and saw the guy. I was still sixteen.

"The recruiter said, 'Well, when will you be seventeen?' I said, 'Next week.' He goes, 'Take these tests. Will your mom sign?' I said, 'Oh, yeah, no doubt about that.' I passed the tests and then forgot to tell my mother. The recruiter showed up at my house at 6:30 on the morning of my seventeenth birthday. I said, 'Oh yeah, by the way, Ma . . .' So I signed everything, she signed me away, and a week later off I went to boot camp in San Diego. It was June of 1981.

"Training was twelve weeks. We'd go to Camp Pendleton for rifle training. Back then they called it RFTD, the Recruit Field Training Depot, where you learned your field training skills. They go up there for their Crucible now. It's all in just one area.

"When I went to recruit training, nobody ever talked about killing; it was pretty much going through the motions of getting a check in a block, okay, we did this, it's done, we did that, it's done. Nobody ever really just sat down and talked. We filed a training schedule, and, if it wasn't on the training schedule, we didn't do it.

"After boot camp they made me 08-11, which was a Cannon-Cocker, the only MOS in the Marine Corps that did not have a school. They took high school dropouts and made them 08-11 because they didn't have to spend the money on them. From boot camp I went home on leave for ten days, then straight to Twentynine Palms for my first unit. I got there November of '81 and I left there in December of '82, went to Okinawa for a year, and checked into Camp Lejeune in North Carolina in January of '84.

"I was at Camp Lejeune until January of '87, when I came down here to drill instructor school at Parris Island. Why did I volunteer? I just idolized my drill instructors. I wanted to make a change in the Marine Corps. When we trained we did a lot of things that just didn't make sense, and I didn't understand. I had a first sergeant who was a Vietnam vet, and he said, Well, you got to go down to where they start at. Don't get out of the Marine Corps. You need to get the rank where you can make changes, but you should also go to where they start by making Marines.

"An example of a waste of time training was something we did called the Marine Corps Combat Readiness Evaluation System, the MCCRES. We did this knowing that it didn't make sense because in combat this stuff ain't gonna work. I would talk to this old first sergeant, who had about five tours in Vietnam, and he said, Well, that's just the way the system is, and you got to stick around if you want to change it. And he would always tell me, Train the way you're going to fight. And that wasn't always a popular decision because I would ask the CO, If we don't fight like that, why are we doing it? Well, that's because the checklist says that. I would say, Well, just because we do it that way don't make no sense. See, when I came in, most of the sergeants and almost all the staff NCOs were Vietnam vets. They didn't so much agree with what we were doing but they weren't involved in training.

"I had trouble getting out of my MOS because we were short and we were deploying so much, but I finally made it to Parris Island in January of '87. I did six cycles as a Green Belt and two as a senior. I was a senior drill instructor as a sergeant, and they're kind of rare. I picked up meritorious staff sergeant right before I left, in January 1990.

"I left Parris Island, went back to Camp Lejeune, and then to the Persian Gulf for Desert Shield/Desert Storm. I was platoon sergeant of an artillery battery, right back in my MOS. The young Marines out there performed super, absolutely spectacular. It was a learning process, because you don't do anything the way you're trained to do it. Specifically, things like convoy ops, riding in the back of an open five-ton truck, you're just a target. So what we did when we sandbagged trucks and all that we found out it was too much weight; we started snapping axles. So we had to work around that.

"In June of '92 I was an instructor over at the staff NCO academy at Camp Geiger, which is part of Camp Lejeune. One thing that stuck in my mind had to do with 'minimum standards.' I always preached that minimum standards was a starting point, and not a goal. The minimum standards were always acceptable but, for me, for sergeant and above, that was unacceptable. I preached that to the sergeants and staff NCOs. I'm talking about physical fitness, appearance, field training, anything measurable. Was that a radical notion? Seemed like it at the time. One CO kept telling me, 'Post, this ain't your battery,' meaning I should mellow out. Me and him never got along too well. And here they're all leaving with Navy Coms (commendations) and Meritorious Service Medals.

"After leaving the NCO academy I picked up gunnery sergeant and went back across the river to Camp Lejeune, where I was the battery gunny in the Headquarters Battery of the Tenth Marine Regiment. As a gunny, by now, you know you start getting set in your ways, and I had a battalion commander by the name of Lt. Col. George Dallas and a sergeant major named Thompson, both still on active duty, and they had a training mindset that really helped me out because they preached, 'Train the way you're going to fight.' And we did. We trained hard, but we didn't do busy work. We trained to be miserable, building the kind of endurance you would need in a combat situation. I really enjoyed my tour there.

"I was gunny when we went to Kosovo and I was with the 26th Marine Expeditionary Unit with BLT [Battalion Landing Team] Three-Eight [Third Battalion, Eighth Marines]. We were the initial entry force in there, the first American forces on the ground. They made my firing battery a provisional rifle company. As a gunny, I had trained them in all the basic rifleman skills, and it paid dividends because we became a rifle company at the drop of a dime, no warning. We were in a couple firefights, patrolling, doing security ops in the little towns, the whole nine yards. And those Marines just performed superbly.

"An artillery battery is a lot smaller than a rifle company. There's only about 130 of us. I was the battery gunny, which is the same as the company gunny. My platoon sergeants, the artillery guys, became rifle platoon sergeants, and what I preached to them was, If you lead, you will win. Leadership, leadership, leadership, no matter what job you are given as a Marine. The platoon sergeants knew all the weapons systems, and it was actually easier on them because we didn't have a million moving parts, like in artillery, to deal with.

"Another thing I always preached was about killing somebody. I always believed the bottom line is, the Marine Corps exists is to kill people. You can filter through all the political rhetoric but the bottom line is, we exist to kill the bad guys. And I preached that to them and I preached that to them, and when it came time to do some killin' for the first time I had to yell at 'em to do it, but they did it, and they did it well.

"We were spread thin. We made up three platoons and we set up road blocks and established security in three different towns. We got kind of in a jam and all I had left with me was the three sergeants, and those three platoons. I never had to look twice. They grew up overnight and they didn't miss a beat. That was their first taste of combat and they just performed superbly. They saw lots of dead bodies. The ones that needed killin', they killed and they didn't murder, you know. We killed people that needed killing. And on the other hand, there was a saying, I don't remember where I heard it, it was so long ago, but it calls the Marine a unique creature. 'He kills with one hand and caresses with the other.' And I got pictures and pictures of all these homeless children from the refugee camps that came back to these towns that were destroyed and their parents were all murdered. And these Marines were taking care of children and puppies on one hand, and killing the bad guys on the other. It was remarkable.

"You had the Kosovo Albanians and then you had the Serbs, who had come through and purged all the Kosovo Albanians. The Serbians started shooting at us and we had to shoot up with them a little bit. We also disarmed several large units of Albanians, the KLA—Kosovo Liberation Army—that were coming down out of the mountains once the Yugoslav army left.

"We were there from June until August. It was pretty serious stuff. We had the whole grunt battalion too, the 26th MEU [Marine Expeditionary Unit]. The battalion landing team was Third Battalion, Eighth Marines. When you go out on the MEU, each battalion has an artillery battery attached. That was where we came in. The rules of engagement were very clear, which was good to see, and I wound up getting a Bronze Star with a V and the sergeants that were with me got Navy Coms with a V, which was very rare, in 1999, let me tell you.

"The thing that really ticked me off in Kosovo was, you know, they called this crap peacekeeping. How do we keep peace? We kill the bad guys. If you act up, we're going to kill you. After what happened in Kosovo when the bad guys shot at my guys, I believe that 99 percent of other units would have let go and just radioed in. But I asked the Marines, I didn't know if any of my guys were hit yet, and I asked them, Can you see them shooting? They said yes. Are they shooting at you? They yelled back to me yes. I said, 'Kill 'em. Kill 'em.' And that's why we wound up doing what we did. Peacekeeping to me is horseshit. It only takes one bullet to end your war, and I'll be damned if it's going to happen to one of my Marines on my watch because of being restrained. And those Marines understood: My God. This ain't peace-keeping. These people are trying to kill us. You turn your back on them for one minute, they will kill you. Damn right.

"We got back from there and I was with the Tenth Marines when I picked up first sergeant and joined an infantry company. Once you pick up first sergeant you can go to any unit. I went across the street to the Third Battalion, Sixth Marines, and we went to Afghanistan with the 26th Marine Expeditionary Unit, BLT Three-Six. We were actually on the ground three or four months until the Army came and relieved us.

"That action was mostly Special Forces guys running around, but again the Marines just performed superbly. You cannot tell they're scared. They focus on taking care of each other. They just are magnificent creatures.

"As for women, I'm not saying they don't belong in the Marine Corps. But men and women are wired different—physically, mentally, and emotionally—we are different. These eighteen-, nineteen-year-olds, they got these things called hormones, and when you put them together with women and you put a little stress on it, things are going to happen. It has nothing to do with professionalism. These people are human beings. If you get in a firefight, I'm not saying a woman ain't going to perform, but what I am saying is a man is going to be more apt to protect the female than to focus on his mission. And there's nothing wrong with that. It's not a lack of professionalism, it's not a lack of training. It is not saying a woman is weak.

"If you take the average man and your average woman, the man can go longer and harder and carry more weight. Men are less disturbed by shoving a knife in somebody's gut or blowing somebody's head off because we are not as emotional, as a rule. Does it bother us? Yes, but usually we can push through. I think men are better at shoving things down because there's a time and place to deal with it and, when you're in a frigging firefight or you're in combat, you cannot deal with the crap. You've got to shove it down. If you let it get to you then, it's going to cause problems. But you've got to deal with it afterwards. I think as a whole women make decisions based more on emotion. That's the way they're wired and it's not good or bad. But that's my opinion. I guess you can put that down. There's nothing wrong with that.

"It's just how we are. I've never been in an integrated unit. Being in artillery and infantry, I didn't have to worry about it, but the ones I worked with, the females at the staff NCO academy when I worked there, the students that came through and the female instructor I worked with understood, and they did well. I told them, You are a lady first. Don't lose sight of that. It's not good or bad, it's just how it is. Now you also got to be a Marine. How do you balance those? I don't know, but a lot of them do it. You hear talk about 'mentoring,' and I cannot stand the term 'mentor.' Because that's just a trusted leader and counselor and it's part of leadership. You lead: It's a whole package. Forget this so-called being a mentor. You're either a leader or a leadee.

"They call it our soft skills. Well, any good leader has compassion, he or she understands that. But the women Marines need to be led by females that are gunnery sergeants, first sergeants, so how do you explain to them how to get there? I can't do that as a man. I can teach them leadership, but walking that fine line on being a lady and a Marine, that's a tough job. The ones that do it, God bless them. We need more of them around.

"There seems to be no middle ground where females are concerned. They're either really, really good, or really bad. And the bad ones make it twice as tough on the others. For instance, if a female gunnery sergeant or a first sergeant checks into a unit, she has to prove herself. When a male gunnery sergeant or first sergeant steps into a unit, it's assumed he knows what he's doing, and, until he screws up, he's really not looked at too hard. But every time a female shows up, she has to prove herself.

"When Marines are not training to go on deployment or anything, they tend to get bored. I think the leadership fails to focus them. Remember: Regardless of what your MOS—the butcher, baker, candlestick maker—the Marine Corps exists to kill people. I don't care if you're turning wrenches on a helicopter or flipping burgers in the chow hall. Either you're actually pulling the trigger or you're doing something to enable the guy pulling the trigger to do it more efficiently. Every Marine is important and *don't* lose sight of that. Your MOS is secondary to the fact that, first of all, you're a Marine contributing to the team to kill the bad guys."

Discussing some of the ways the Marines have changed since he enlisted, Sgt. Major Post said, "I think we've gotten better where we've stopped glamorizing alcohol, and I think that was a good move because it avoids a lot of problems. You know there's no problem you can have that alcohol ain't going to make worse. We pretty much stopped that. Marines realized you don't have to be a drunken thug to be a tough guy. I'll tell you the toughest Marines I've ever met are the nicest guys in the world.

"We take care of business behind the scenes, not to look good but to remind ourselves that, when the shit hits the fan, it happens quick. Combat is 99 percent boredom and 1 percent uncontrolled chaos and violence.

"As for Iraq, a lot of my friends, first sergeants and sergeants major, are over there, and they report the Marines are just performing superbly. Nearly all Marines are chomping at the bit to get over there. You don't hear them complaining about deployment time. This is what they came in the Marine Corps to do. They're warriors, dealing with snipers and IEDs, Improvised Explosive Devices; I'm old school. I call them booby traps, because that's what they are. Mostly, it's very frustrating, but they're doing their job wonderfully, and as usual 99.9 percent of the stuff that's going on don't make the papers, just the bad stuff. The people over there absolutely love them.

"These guys are warriors now, and the thing is they fight for their own tribe. Push comes to shove, they're fighting for their fire team and the squad that they're with. If you screw with one of the people or the people they're in charge of, they're going to clean you out. And that's what they're doing.

"You always hear about how much tougher the old guys were in the old days, but it is NOT true. When we went to Afghanistan, these Marines were carrying so much weight, at least seventy-five pounds per man, that they could barely walk. Marines today look different, they have different uniforms, they got different weapons, but they fight like any Marine in history. And that's what these guys talk about:

They talk about the Beirut vets, the Gulf War vets, the Vietnam vets, the Korea vets, the World War II vets, the World War I vets. And they love it.

"As for me, I'll do like I have been, do what they tell me and take Motrin and suck it up and keep on plugging. I hope to get an infantry battalion soon. You know, when I got two years in down here, I hope to leave for an infantry battalion and get to Iraq, get back in the fight. And not because I'm a warmonger; it's because I know what to expect. I've been there three times, I've been in combat three times, I know what to expect. And if I can help keep one Marine alive, or help one Marine deal with the stuff that comes after, then that makes it all worthwhile. They need to know it's normal to be scared.

"And I like to remind them, Remember, you're writing the next chapter."

CHAPTER TWENTY-EIGHT

KEITH BURKEPILE

Director, Drill Instructor School
Parris Island
1985–

You know, Chesty Puller said something
to the effect that, if our nation continues to be this soft,
a foreign soldiery will someday come and take our women and children
and breed a hardier race. That's not word for word.

Major Keith Burkepile, director of drill instructor school, Parris Island. "You don't run from your weaknesses. You attack them." He said drill school students "do not come here for $375 extra a month or for meritorious promotion. They come here to make an impact on our Corps and to be like their drill instructors." (*Photograph by Cpl. Brian Kester, Combat Correspondent, MCRD Parris Island*)

I spoke over the phone for a long time with Major Burkepile (pronounced Burkeypile) as he reviewed in detail the challenges faced by drill instructors before they ever even see a recruit. He also compared training from his day, the mid-1980s, to the present, 2005, and did not find it was all that different, despite what appear to be easier aspects of training today. He said, "You always hear that statement, 'Back in the Old Corps.' I still have no idea what that is. I don't know where the cutoff is, where the 'Old Corps' ended, if it's World War II, or before." He said today's young men and women were not as fit as they were in his generation but that, overall, they were smarter. He also had some interesting observations about the work ethic of today's recruits. He was extremely courteous, and patient. The transcript of my interview with him ran to twenty-six pages, single-spaced.

"For every Marine that comes here to drill instructor school, even the biggest studs, their bodies are going to break down the first month. There's just no way. Because they're just not used to it. The PT program is no joke. Each session is two hours long and it's a legitimate two hours, straight through. They may have a high PFT score when they come here, but PFT is only one little indicator of physical fitness. We do a lot of obstacle work to prepare them for conducting obstacles, a lot of upper-body work, and of course we focus on their endurance in both hiking and combat-type of endurance and get their speed up, you know, work their short-twinge muscles on the speed side of the house so their PFT run time will improve as well, so when they get out there as a Hat, they can lead a squad-ability group of fast recruits. They have to master every obstacle while they're here on the Obstacle and Confidence Courses. That's a requirement. They have to achieve a first-class Physical Fitness Test to graduate.

"The physical conditioning program here is one of the tougher ones in the Marine Corps. It certainly will be most challenging for a lot of these students. When I say that I'll caveat: There's other schools out there, even other units that have tougher PT programs than we do. However, ours is designed to prepare them to work a sixteen-, eighteen-hour day, for them to be on their feet all day long, and to put them in a position where every time they step in front of those recruits, no matter what that recruit's ability is, they're going to be able to lead that recruit by example in everything they do, whether it be running, going out to the Confidence Course, or the Obstacle Course, and demonstrating obstacles in front of those recruits. Recruits need to see that drill instructor as perfect, even though he or she is not.

"The longest run they'll do here is five miles. I will tell you that the toughest run they'll do here is called Iron Mike [see Chapter 2]. What it is, you run in boots and utilities, and we run four miles, total, boots and utes. We run to the five Obstacle Courses on the Depot. We'll run the First Battalion O Course, execute that, then start running again, to the Third Battalion O Course, execute that, hit Support Battalion, hit Second Battalion, hit Fourth Battalion, and then come home. That's a break-all for anybody. Everything that we do here in PT, myself and all the instructors are doing every bit of beside the students. Everything's leadership by example.

"I've done four marathons, but I'll be honest with you: I'm a horrible runner. A lot of people think, since I'm the director over here and I'm constantly running, that I'm good at it. I have a horrible stride and my left leg's shorter than the right. If the Marine Corps knew it, I never would have got in. I wear an insert in my left heel and, because of it, my stride's very inefficient, too much vertical, not enough horizontal, but you know what, you don't run from your weaknesses. You attack them. Every time I've completed a marathon, even though I've never met my goal time, it's still always great to be standing there at the finish line with everyone else. Right now my focus is on trying to be fast enough to keep up with the students who come here.

"Men and women train side by side at drill school. We don't choose them. We are still an all-volunteer force on the drill field. They're mostly sergeants and staff sergeants. The number of drill instructors on the base fluctuates between 400 and 500. We graduate about 200 a year. There were 52 in our last class. Training lasts eleven and a half weeks. The instruction program for the school is 400 pages long. Our instructors consist of eight gunnery sergeants and then I have a first sergeant who also is the chief instructor of the school, dual-hatted as a first sergeant/chief instructor. All are noncoms, nomcomissioned officers, except for the assistant director, who is a captain. I came over here as director of the drill instructor school in March of 2004 after arriving as the battalion executive officer the summer before.

"Of course, drill instructor school is far more challenging than recruit training, for many reasons. All the students here are already United States Marines, so they have the ability to perform and they're proven leaders. They are not the top 10 percent of NCOs and SNCOs, but all are proven in their MOSs. They do not come here for $375 extra a month or for meritorious promotion. They come here to make an impact on our Corps and to be like their D.I.s. My instructors put a lot of extra time into getting them across the stage at graduation—just like D.I.s will with recruits. A lot of time and effort are invested above and beyond what is called for in the instruction program.

"Extra hours are spent after our normal day of '0500-1900' working with students who are having problems—much the same as the D.I.s. We probably average 90 hours a week here at the school. A D.I. averages 110–120 hours a week, especially since he or she sleeps with the recruits every third night. We do not sleep with our D.I. school students as obviously they are United States Marines and do not require that much supervision.

"Students are required to memorize 'verbatim' and teach back while demonstrating twenty-three different drill movements. The final book contains unit drill movements that are over four pages long, single-spaced. Sometimes we get Marines with a third- or fourth-grade reading level, and then we have to get that Marine to perform these teach-backs, so he or she can pass the school and be confident in his or her abilities when they stand in front of eighty recruits to teach drill. We need three successful years out of every drill instructor.

"I had a student fail Book 2 [Individual Movements without Arms II] for the third time this morning. Unfortunately, he will not graduate. He is a good Marine, but not cut out to be a D.I., once again proving that every Marine has the potential to be a D.I., but not every Marine will or can be a D.I.

"The Standard Operating Procedure for recruit training is a big document, and there's no way any Marine can memorize every part of it. But we do test on certain parts they need to know, certain parts they must have in their heads, whether it be what can you do under certain flag, or temperature, conditions, when lightning threatens, and so on. But there's also a little pocket guide, a chapter out of the SOP they have to keep with them at all times. So you have the important parts in your head but you know where to look if you have to find other information quickly.

"The women do a good job for the most part, but if anyone's going to say that the male body and the female body are the same, I would call them a liar. It's just not a fact of life. It's not the way we're made. For instance, the obstacles on the course at Fourth Battalion, the female battalion, are lower. Certain obstacles in the Confidence Course they either don't have to do or don't have to do as much of as the males. They do not have to do pull-ups. They do the flexed-arm hang, which is not the same. In our point system, they have a three-minute advantage when they start a run. To max a three-mile run, a male has to do it in eighteen minutes or less. A female's got to do it in twenty-one.

"Obviously, the official Marine Corps stance would not say this to you, acknowledge the differences, but I will say I'm good with that [recognizing the differences], because if we don't recognize that, we're just being dumb. So don't stand up and say that males and females are trained to the exact same standard. But I will tell you this: The females in this school are right beside the males.

"They're in the same formation on the hikes and in the run formations. We did our mid-cycle Physical Fitness Test yesterday. I have nine females on deck right now, and if you subtract just one of them, the other eight just did phenomenal. One of them got under twenty-one in her run, and the others were all under twenty-four minutes. Most were right around that twenty-two-minute mark. They're doing great. They're in the fight. And they get treated the same as the males.

"So you can recognize their physical differences, but female Marines just need to be treated like United States Marines, not like females. And that's how they're treated here, the same as the males.

"However, I do believe segregated training for recruits is more effective than the approach taken by the Army. I just think there's too many problems if you don't separate them. The drill instructors are doing everything with these kids. When they're here, not only are they a drill instructor and their mentor, but they're basically like a father or mother. I mean, they're putting these kids in the shower. Some of them still don't know how to hygiene properly. On the male side, they're teaching them how to shave.

"Hygiene is a huge deal here at Parris Island. We have a lot of problems, especially in the summer, with cellulitis and a thing called MRSA, which in layman's terms is a microorganism that gets into the skin and can cause infections, boils, or pneumonia. If recruits get that, a lot of them end up going home. Parris Island is a germ factory, especially in the squad bays, and that's why hygiene is just unbelievably critical. Once an infection gets in a platoon where you're living amongst each other every hour of the day, you can't get rid of it. You go to sick call and guess who else has got it? The Hats.

"There's no such thing as a sick day for a drill instructor. He can't call in sick. It doesn't matter. You got the crud, you got a fever, you're working. And you will not let the recruits know that you're sick. My entire last class I could not get them to get over the crud. It was ridiculous. Tell them to take vitamins, take their medicine. All you can do is fight through it.

"But I just don't think, if I'm a parent and send my daughter here, I don't want a male Hat putting her in the shower with a bunch of other females. And I wouldn't want to put that male drill instructor in that position. I don't think it makes sense. I don't know how the Army does it. And do I really want to put males and females around each other where embarrassing situations could occur? A female can't do something that a male can, a male can't do something that a female can. Either way, it's not good.

"I grew up in the Carlisle, Pennsylvania, area and joined the Marine Corps in November of 1985, and went through Parris Island as an enlisted man on a ninety-day Reservist program, delayed entry, in June of 1986. I was a good recruit, but boot camp was just so tough. I mean, I was a young kid, I was seventeen, I weighed a buck twenty, I was five feet seven. I was a baby still. Drill instructors never had any problems getting me to open up my mouth or move as fast as I could. I did everything as hard as I could. I wanted to be a Marine but I will tell you it was hard for me. When I went to my first Marine Corps ball, I was the youngest Marine there. I had just turned eighteen.

"I went home after graduating from boot camp September 3 and started working and went to college part time. The following summer I went through infantry training school at Camp Lejeune as an 0311 infantryman. I did volunteer but didn't know what I was doing. My recruiter didn't really explain to me what an 0311 is, what an infantry-man does. I didn't even know what an MOS was. In fact, my drill instructors asked me when I was here at Parris Island and I said, 'Sir, this recruit does not know what an MOS is, sir.' They sort of laughed a little bit. One looked at my book and told me, then laughed even more. So I just sort of stumbled into it and it was just ironic I ended up liking it. Pure luck.

"I truly didn't get my love for the infantry until we did an active-duty stint just prior to Panama, with Third Battalion, Sixth Marines. I was a squad leader at the time, integrated in an active-duty platoon. We filled in Table of Organization shortfalls with India Company. We just filled in with the active Marines and I ended up being third squad leader. Then we started training in the jungles of Panama, and that was an eye opener.

"Our platoon commander, Nicholas Place, was one of those guys who never cared about getting promoted. He only wanted to command a platoon. He had actually been in the British Royal Marines. He was an American citizen by this time, now in the U.S. Marine Corps, but he brought so much to the table, the professionalism we may have been lacking a little bit.

"When he came and we got activated for Desert Shield/Desert Storm, we ended up not doing anything, but we just trained and trained and trained our butts off constantly in preparation. This was in 1990, '91. We were actually out at Pendleton. We were being told you're going over there to Iraq, yadda-yadda. But we were living a life that I wish I could always do with Marines, ever since I've been an active-duty officer.

"We would go to the field on Monday, train all week, and then come back Friday, make sure our gear was clean and our rooms were tight, and all the good things that Marines should be doing, and then we'd get a little bit of liberty. Then we'd start it all over again on Monday. Every week was a new thing that we needed to be up to speed on in order to be ready to go to combat. Of course, we never ended up going because the war ended up being shorter than what they thought. But we got to the point where we were proficient in every area of our job.

"I had wanted to be a grunt, but, because of the cutbacks that had been made after Desert Storm, they didn't need any sergeants or even corporals on active duty. I went through OCS in January of '94 at Quantico, got commissioned in March, switched to artillery, and went straight to the Basic School. My first duty station as an officer was Twentynine Palms, California, I had obtained my civil engineering degree from West Virginia University while I was a Reservist. My math background helped me on the gunnery side of the house in artillery.

"I had wanted to stay in the infantry, but the bottom line was I didn't earn it. During the straw pools, I'd gotten infantry every time. I thought, I'm good. On the actual MOS draw day, I had actually moved up. I'm sitting in the middle portion of the top third of the company in academic standing, like forty-fifth, right in the middle. One of my good friends at the time was standing right in front of me. He went in and put his little name on the peg board for the last infantry slot, and I'm like, I'll be damned. So I looked up at artillery, my second choice, and put my name on one of the four pegs still vacant. For a while I was certainly bitter. I remember one of the captains looked at me and he said, 'Burkepile, what is wrong with you?' I said, 'Aw, sir, I didn't get infantry.' He said, 'Well, what did you get?' I said, 'Artillery.' He said, 'Well, you're still killing stuff. You're just further away when you're doing it.' You know, that got me to put a little pep in my step at the time, but I was sort of bitter for a little. Yet, in the long run and personally in my own beliefs, I do believe that God has a plan for everybody, and that's just me. But I believe there's things happen for a reason.

"Artillery ultimately ended up being an awesome fit for me. There were so many things I got to do as an artillery officer that I wouldn't have got to do as an infantry officer, and, besides, we serve right beside the infantry. And I got out there as a forward observer. I was attached to an infantry company. I also ended up being an FO with a tank company. You're right with a line company. And I served with a great infantry company. And artillery always has a second mission, of provisional infantry. I was able to teach a lot of that to my Marines in my platoon.

"It's funny how I got back to Parris Island. Part of me always wanted to come back here just because when I left here one of the things I said was, I never am going back to that place. I was always told about the sand fleas by my dad's buddies that were in Vietnam, and of course when you get here now . . . you know, now they carry Skin So Soft in the recruit PX. When I got down here you had to write a letter home to mom and say, Please send some Skin So Soft, because that was one of the best things to keep the sand fleas away. That, combined with bug spray. We're outside all the time at the school, and I put a layer of Skin So Soft on, cover it in Off bug spray, and I'm good. Keeps them away from me. Everybody's got their own little remedies. A lot of my instructors mix Skin So Soft and rubbing alcohol and bug juice and put that on. It works for them.

"But there was a time when I thought, I'm never going back there. As much as I looked back fondly on the experiences I had here with my platoon and as much as I worshipped my drill instructors, I was like, Nah, I ain't going back to Parris Island. My whole association with Parris Island was as a recruit. A lot of times in life you don't appreciate things until later anyway. You just don't. So the young kid side of me thought, God, that is just amazingly tough work.

"Eventually I thought I wouldn't mind going back as an officer. Then I heard the horror stories about how officers were basically acting as spies against the drill instructors. Series officers. That's just me hearing it, not me being here seeing how things are done, just hearing it from maybe former drill instructors. I have no desire to be a spy because I have always had a certain affection for enlisted Marines and I love serving beside them.

"Then, when I became a major, I thought, maybe you can go back there, and have an impact. I got married in January of 2004, before I came down here. I played the single life until I felt I was done with it. I have an eight-year-old, Marcus, I call my child; he's a stepson. My wife, Kimberly, and I also have one on the way. She's due in June.

"Now, I'm expecting to be at Parris Island until the summer of 2006. How long I remain as director is up to the commanding general and the commanding officer of the Recruit Training Regiment. There's no guarantees. If they decide they want to put another major in this job and pull me, it could happen tomorrow. I will say that being the director of this school has been one of the most rewarding jobs that I have held. It's an awesome job, that's for sure. I love it. I get to work with the best Marines.

"But I'll tell you, I've been in a learning process since Day One, learning what it takes to be a drill instuctor, to actually make a United States Marine, and, more important for me, what is the officer's role in that? It's hard work to come down here and you truly, as an officer, need to be able to lead these drill instructors. God bless all the Marines, the military, in the fight in Iraq, Afghanistan, Horn of Africa, wherever—they're obviously the true warriors on the tip of the spear—but I will tell you that being at Parris Island may be one of the toughest jobs for a Marine to come and do, period.

"I'm not going to lie to you and tell you every officer's doing eighteen- to twenty-hour days, because they're not. When I first got here I was feeling things out as an XO [Executive Officer], trying to figure out where the officers fit in. You talk to sergeants major, you start feeling out what you need to do. It took me a while to understand it, but now that I'm here there's so much common sense to it. If I'm out in the operational force and I've got a training schedule and I know my Marines are going to be on deck at zero five, and they're going to have final formation at 1800, I'm not going to show up for work at zero eight and leave at 1630. I'm not going to do that. I'm going to be there fifteen minutes ahead of time and I'm going to leave after they leave. That's been going on for years. The drill instructors aren't going to work lights-to-lights every day either. They would never make it. They are going to work sixteen hours almost every day. In my opinion the officers have at least got to make an effort. It should not be any different than it is out in the operational forces. It's not spying. You need to be there to support your Marines.

"It's still a sergeant or staff sergeant and a captain or a lieutenant. They're still writing a fitness report on that Marine they're responsible for. The problem is that a lot of the young officers will get intimidated by the drill instructors. It's amazing what happens when a Marine puts a campaign cover on his or her head, and a belt around the waist.

"The officers go through a series officer course here, but it's only two weeks. They get a quick firehose of what they need to know to go out there and serve, but they don't need it. What they need to do is take the basic leadership principles that they've used in any other unit they've been in. They still apply. It does take a lot of time. If I'm a drill instructor out there and I see the captain show up at 0700 and I've already done been to the chow hall with the recruits, got cleanup done, and now he shows up and then he goes in his office and sits and I don't ever see him and he comes out once in a while and maybe he wants to correct me on something and then he goes home to Mama at 1700 and I'm still in here burnin' and churnin,' I personally don't have a lot of respect for that.

"The drill instructor's role is to train the recruit. The officer's role is to insure the integrity of recruit training, to insure that every United States Marine is trained to the same standard in accordance with the Standard Operating Procedure. A lot of the officers are lieutenants when then first get here and then they'll get promoted to captain not long after. Here's a tough thing for them: Their job can be boring. They're watching, constantly standing back, but they're needed because a drill instructor's in the fight with these kids sixteen, eighteen hours a day. A schedule like that for an extended period wears you out, mentally and emotionally.

"Some of these kids that come here are going to be what I'd call a punk. They didn't sign up for this; they did, but they really didn't. They didn't know. And now they get here and they don't want to play. So they will give the drill instructors a hard time. Just basically refuse to train: 'Nah, I ain't doing that.' They're rolling their eyes, they're being belligerent. Some of the drill instructors' biggest challenge is to be patient, to be mature, using the tools available, the incentive training techniques and their voices. But a drill instructor's going to make mistakes. We're human.

"And the officer's got to be separated from that emotional struggle and have the moral courage, if a drill instructor makes a mistake, to say, 'Time out. Hey, drill instructor, I just need to talk to you a second. Can you come over here? Let the recruits stand and come over here.' That's when you have a man-to-man conversation with the drill instructor. The series officer should be doing this.

"The series gunnery sergeants serve right beside the series commanders. Series gunnery sergeant is responsible for insuring that those Hats are doing the right thing, that everything is being done in accordance with the SOP. The series gunnery sergeants also lead PT sessions.

"There are four companies in each of the three male training battalions. The women's battalion has three. There are two series commanders per company, twenty-four series commanders on the male side of the house and six series commanders in Fourth Battalion for the females. They're going to have two series, what we call lead and follow series. All the series commanders are coming off at least one tour in operational forces.

"The drill school graduates start as the third or fourth drill instructor in a training platoon. You're first going to be what they call an 'experienced' drill instructor. Back in the Day it used to just be called a Heavy. The Heavy teaches the recruits everything. The senior sort of plays a role being Dad. The senior takes care of the drill instructors, insures the recruits are being taken care of, insures that things are being done properly. The 'experienced' drill instructor is teaching them drill, teaching them basic daily routine.

"We only give them the basic tools here at the school, so when they first get out there, their goal is just a firehose of learning. They're going to be in good shape when they leave here, know the SOP, know drill, know general military subjects, all you need to reinforce the recruits. Their uniforms are going to be tight. It's going to take them one or two cycles to learn how the training schedule works, how the basic daily routine works.

"Recruits are constantly reinforced with Marine Corps history. They get classes in academics, but the drill instructor's going to break everything down into very short things recruits can repeat, so, while they're out there hiking, they'll start spitting out knowledge about the five major battles in World War II. The recruit will repeat, 'Sir, the five major battles in World War II are . . .' and they'll start spitting them out: Iwo Jima, Okinawa, Guadalcanal, Bougainville, Tarawa. History is always important to an organization. If you don't know how you got to where you are, how can you appreciate how you got there?

"If I can't appreciate what every Marine that served in our Corps did in order for us to even be here today, I don't appreciate what I have right now. It doesn't matter if you look at just the basics of Guadalcanal or, for instance, I was reading a little blurb on Hue City in Vietnam, where 2,500 Marines attacked and defeated 11,000 North Vietnamese. Now when you go into the offense in most cases, you want a three to one advantage. In Hue City, it was reversed. I mean, how can you live up to that? All of us get to the point where we want to whine or complain about the long hours at work or, screw this, I'm not doing this any more, I'm ready to get out. But if you stop and think about stuff like that, it'll make you appreciate how easy you got it.

"You've got to believe in what you're doing. One of our biggest challenges is getting recruits to believe. In the civilian world, they would call us anal retentives. The little things that we make a big deal about here—even Marines out in the operational forces would be like, You gotta be kidding me. Get a life. But you have to have that attention to detail. Our job is to teach everything, not 50 percent of it, not 75, but 100 percent of it right.

"We have Marines coming to the school from every different unit, every MOS, and a lot of them when they get here are like, 'No, no, I thought I wanted to be a drill instructor, but I really don't. I don't want to work this many hours a day and I don't want to do this.'

"The dropout rate changes from class to class. The male attrition for recruits every year seems to be about 9 or 10 percent. But my attrition rate at the drill school varies from 10 to 30 percent. Most of my attrition is not due to Marines who fail. It's due to Marines who come here with prior injuries that weren't fully healed or identified. They're supposed to be screened by a doctor and signed off on. But right now, in this class, I have disenrolled five Marines, and four of the five are due to prior existing injuries. One of them showed up on crutches.

"Training today is just as tough as when I went through. There was a time when I didn't think that, but I was ignorant. It's just as tough and it's very much the same. The SOP may be a little thicker than in 1986 but what I didn't know about my Hats, the D.I.s I had, was that they were very professional. You always hear that statement, 'Back in the 'Old Corps.' I still have no idea what that is. I don't know where the cutoff is, where the 'Old Corps' ended, if it's World War II, or before.

"Most of the differences today are on the PT side. It's smarter. For instance, now they have this program where they rotate in the beginning, in what they call Go-Fasters here but are really sneakers, and boots. When I was here it was just straight up, Get your boots on, let's go. I didn't have problems with stress fractures. But you talk to civilians who work here on the athletic side and they'll say today's kids don't drink enough milk and their bones aren't as solid and this and that, but my point is you can't run in boots and do everything in boots because if you ramp up, you are going to break some kids unnecessarily. They're not fit.

"A lot of schools don't have gym classes now. I have to all but kick my son out of the house. He'd rather stay inside and play Nintendo or watch television. When I was a kid we were outside all the time. But these kids come here now and a lot of them have have never done PT one day in their life. They never even went on one run.

"We've taken out all the running in boots and utilities, for the most part. If you talk to the civilian athletic trainers, they say studies show the average soldier or Marine in combat will never run much further than 500 meters at a pop. That may be true, but a lot of the studies are based off Desert Shield or Desert Storm and we were mecced up for a lot of that in AVs and stuff, or using helicopters to get in and out. But what if we have to go fight a dismounted fight in another Vietnam-type country or in a Korea? In Korea, mechanized vehicles can't go everywhere. I know because I've been there.

"My point is if you get used to running in boots and you do get used to it, your body gets used to it. We still do boots and utes runs in D.I.

school and I'm not fast and, yes, I'm getting older, but I will run in boots and utes once a week. Why? Because it's a mental thing, it's mental confidence, they're heavier and they're what I wear all day long. I don't wear sneakers all day long. I'll be thirty-seven in October. You know physically it's a young man's game.

"So some of that's changed. I did everything in boots when I was a recruit. But the stress that's on the recruits, the intensity they're trained at, what they gotta do, from pugil sticks to martial arts, to the Confidence, Obstacle Courses to the knowledge they got to bring, the history they got to learn, the drill they execute—it's all just as tough. As for the talent pool, I would say that intelligence wise, book smart, what they can grasp, yes sir, they're good, but I will tell you right now their physical conditions are a challenge.

"The work ethic may not necessarily be there now too, so you're teaching a kid not only how to be a U.S, Marine but teaching a totally new mindset. If they've never been held to the fire in anything they've done and they have been handed everything, they haven't worked for anything. I think kids want to give up easier now.

"Of all the ones I started recruit training with, only a couple dropped out. And once we started truly training there wasn't any one of us that would have given up. You'd have had to shoot us in the head. And now I constantly see kids who say, 'I want to go home.' Why? Don't you want to be a U.S. Marine? 'No, sir, I don't.' Well, didn't you want to be a U.S. Marine when you signed up? 'Well, yes, sir, I did.' Why don't you now? 'I don't like this. I don't have any desire to do this ever.' And you think, how can you be good with that? I could never have went home and looked my father and mother in the eye or any of my friends and said it just wasn't for me. I gave up. But that happens a lot. And what's amazing about the drill instructors today is the extra effort they put in to overcome that.

"When I talk about stuff like that, Kimberly looks at me like I got something growing out of my forehead. But I just think it's a shame that right now the world is a tough place, even though as Americans we have more than most countries. If you go to any of these third world countries and look at how people have it, how they live, the kids, the families and what they got. We take so much for granted as

Americans and we don't set our kids up for success by preparing them for the real world, which is brutal. You don't get stuff handed to you out in the world and it just amazes me that we're willing to put them out there like this and eventually some day, All right you're on your own now. Well, what have you done to prepare them? You know, Chesty Puller said something to the effect that, if our nation continues to be this soft, a foreign soldiery will someday come and take our women and children and breed a hardier race. That's not word for word." (What Puller actually said was: "Our country won't go on forever, if we stay as soft as we are now. There won't be any America because some foreign soldiery will invade us and take our women and breed a hardier race.")

"The cause? I just think it's the liberalism throughout our society. And the media is a big killer in how people look at stuff. I also think we have no appreciation of our history, not just as a Marine Corps, but the country as a whole. Unfortunately, a lot of Americans are against the Marine Corps because they don't know about it. They're ignorant of what we're trying to do. For example, I think our young Marines are doing awesome in Iraq, and this is a sore subject with me because it's a constant guilt trip. I want to be there. The Marines over there are doing phenomenal. You read about good stuff that's going on in every unit. There's Pfc.s and lance corporals going who are doing great things all the time. So I truly believe the system works. We can get better. I won't lie to you. We can always get better."

Men Counting Cadence

As called by Sergeant Major Bill Paxton (USMC Ret.)
Sunday, September 11, 2005

Platoon, ten hut!

Right face! Forward, march!

Leyo, right-left

Leyo, right-left

Stomach in, chest out, hold your shoulders
    back, dig your heels in!

Let me hear 'em crack:

Strut! Strut! Strut!

Three to the front, six to the rear,

Line, cover! Line, cover!

Settle down, girls, you're bouncin'!

Leyo, right-left

Leyo, right-left!

Change step, march!

Lef', right-left

Delayed cadence

Count! One, a little better!

Two, a little louder!

Three, all together!

Four, that's better!

One, two, three, four!

One, two, three, four!

We love the Marine Corps!

Strut! Strut! Strut!

Lookin' good, lookin' good,

Settle down, settle down

Adle, left, right, left, right, left, right [tune
    of Marines' hymn]

Adle, left, right, left, right, left! Adle,
    left, right, left, right, left!

Adle, left, right, left, right, left!

[repeats three times]
Platoon, halt!
About face!
Freeze! Don't move one single
    solitary muscle!
You just passed the most ultimate,
With your self-discipline,
    your motivation
And your dedication
To be one of the finest.
You have now earned the title
Of United States Marines!
Platoon 269, Dismissed!
(Aye, aye, sir!)
Oorah!

THE KNOWN MARINE

The table is set for the Fallen Comrade at the annual reunion dinner of drill instructors at Parris Island. "I have a young Marine enter the room," said Vic Ditchkoff, president of the National Association of Drill Instructors, "and this symbolizes the presence of the Marine who is no longer with us. On the table is the folded flag, the campaign cover, the dog tags, the Purple Heart, the Silver Star, the inverted dinner plate, and the upturned wine glass. And there are the crossed swords, the rose in the vase, and the candle, representing eternal light." Ditchkoff adds, "Parris Island is a religion. I wish you following seas and Semper Fi." *(Photograph by Larry Smith)*

About the Author

A veteran newspaper and magazine writer-editor, Larry Smith has written on military matters for *Parade* magazine. He also is the author of *Beyond Glory: Medal of Honor Heroes in Their Own Words* and a novel, *The Original*. He is a 1962 graduate of the University of Michigan. Smith, who started out on a Michigan farm, worked on newspapers in Wyoming; California; and Westchester County, New York. He was with the *New York Daily News* and *The New York Times* prior to serving nineteen years as the managing editor of *Parade*. He is a former president of the Overseas Press Club of America and a member of the Explorers Club. While he never attempted basic training in a military unit, he has run four New York City marathons and climbed Mount McKinley in Alaska and Mont Blanc in France. He and his wife, Dorothea, have three children.